Coffee Wilt Disease

Coffee Wilt Disease

Edited by
Julie Flood
CABI
Bakeham Lane
Egham
Surrey
TW20 9TY
UK

www.cabi.org

CABI is a trading name of CAB International

CABI Head Office
Nosworthy Way
Wallingford
Oxfordshire OX10 8DE
UK

Tel: +44 (0)1491 832111
Fax: +44 (0)1491 833508
Email: cabi@cabi.org
Web site: www.cabi.org

CABI North American Office
875 Massachusetts Avenue
7th Floor
Cambridge, MA 02139
USA

Tel: +1 617 395 4056
Fax: +1 617 354 6875
Email: cabi-nao@cabi.org

ISBN-13: 978-1-84593-641-9

Typeset by MTC, Manila, The Philippines
Printed and bound in the UK by the MPG Books Group

Contents

The colour plate section can be found after page 84.

Contributors

S. Admasu, *Jimma Agricultural Research Center, Ethiopian Institute of Agricultural Research, PO Box 192, Jimma, Ethiopia*
C. Agwanda, *CABI Africa, UN Avenue, ICRAF Complex, PO Box 633-00621, Nairobi, Kenya. E-mail: c.agwanda@cabi.org*
P. Aluka, *Coffee Research Centre, PO Box 185, Mukono, Uganda. E-mail: cori@africaonline.co.ug*
Z. Arega, *Jimma Agricultural Research Center, Ethiopian Institute of Agricultural Research, PO Box 192, Jimma, Ethiopia.*
D. Bieysse, *Centre de Coopération Internationale en Recherche Agronomique pour le Développement (CIRAD), TA41/K , Campus International Baillarguet, 34398 Montpellier, Cedex 5, France.*
A. Girma, *Jimma Agricultural Research Center, Ethiopian Institute of Agricultural Research, PO Box 192, Jimma, Ethiopia. E-mail: girma.adugna@yahoo.com*
G. J. Hakiza, *Coffee Research Centre, PO Box 185, Mukono, Uganda. E-mail: cori@africaonline.co.ug*
H. Hindorf, *INRES-Phytomedizin, University of Bonn, Nussallee 9, 53115 Bonn, Germany.*
C. Jefuka, *Jimma Agricultural Research Center, Ethiopian Institute of Agricultural Research, PO Box 192, Jimma, Ethiopia.*
A. Kalonji-Mbuyi, *Université de Kinshasa, BP 117, Kinshasa XI, DRC and Centre Régional d'Etudes Nucléaires de Kinshasa (CREN-K), BP868, Kinshasa XI, DRC. E-mail: adrienkalonji@yahoo.fr*
A. Kangire, *Coffee Research Centre, PO Box 185, Mukono, Uganda. E-mail: cori@africaonline.co.ug*
D. L. Kilambo, *Tanzania Coffee Research Institute (TaCRI), Lyamungu, PO Box 3004, Moshi, Tanzania. E-mail: www.tacri.org*
M. Kimani, *CABI Africa, UN Avenue, ICRAF Complex, PO Box 633-00621, Nairobi, Kenya. E-Mail: m.kimani@cabi.org*
P. Kucel, *Coffee Research Centre, PO Box 185, Mukono, Uganda. E-email: cori@africaonline.co.ug*
D.T. Kyetere, *Coffee Research Centre, P.O.Box 185, Mukono, Uganda, email: cori@africaonline.co.ug*
P. Lepoint, *Unité de Phytopathologie, Université Catholique de Louvain, Croix du Sud 2/3, B-1348 Louvain-la-Neuve, Belgium.*
G. Lukwago, *National Agricultural Research Organization, PO Box 421, Kabale, Uganda.*
H. M. M. Maraite, *Unité de Phytopathologie, Université Catholique de Louvain, Croix du Sud 2/3, B-1348 Louvain-la-Neuve, Belgium.*
L. Masumbuko, *Tanzania Coffee Research Institute (TaCRI), Lyamungu, PO Box 3004, Moshi, Tanzania. E-mail: www.tacri.org*
S. Mdemu, *Tanzania Coffee Research Institute (TaCRI), Lyamungu, PO Box 3004, Moshi, Tanzania.*
A. Million, *Jimma Agricultural Research Center, Ethiopian Institute of Agricultural Research ,PO Box 192, Jimma, Ethiopia. E-mail: girma.adugna@yahoo.com*
R. O. Musebe, *CABI Africa, UN Avenue, ICRAF Complex, PO Box 633-00621, Nairobi, Kenya. E-mail: r.musebe@cabi.org*
P. C. Musoli, *Coffee Research Centre, PO Box 185, Mukono, Uganda. E-mail: cori@africaonline.co.ug*
E. Negussie, *CABI Africa, UN Avenue, ICRAF Complex, PO Box 633-00621, Nairobi, Kenya.*
N. M. Ng'homa, *Tanzania Coffee Research Institute (TaCRI), Lyamungu, PO Box 3004, Moshi, Tanzania. E-mail: www.tacri.org*

J. Njuki, *CABI Africa, UN Avenue, ICRAF Complex, PO Box 633-00621, Nairobi, Kenya. Present address: ILRI, PO Box 30709, Nairobi 00100, Kenya.*
G. Oduor, *CABI Africa, UN Avenue, ICRAF Complex, PO Box 633-00621, Nairobi, Kenya.*
J. Ogwang, *Coffee Research Centre, PO Box 185, Mukono, Uganda. E-mail: cori@africaonline.co.ug*
N. Phiri, *CABI Africa, UN Avenue, ICRAF Complex, PO Box 633-00621, Nairobi, Kenya. E-mail: n.phiri@cabi.org.*
F. Pinard, *Centre de Coopération Internationale en Recherche Agronomique pour le Développement (CIRAD), TA41/K, Campus International Baillarguet, 34398 Montpellier, Cedex 5, France.*
M. A. Rutherford, *CAB International (CABI), Europe-UK, Bakeham Lane, Egham, Surrey, TW20 9TY, UK. E-mail: m.rutherford@cabi.org*
N.T. Saiba, *Office National du Café, BP 8931, Kinshasa Kingabwa, DRC.*
A. Shibru, *Jimma Agricultural Research Center, Ethiopian Institute of Agricultural Research, PO Box 192, Jimma, Ethiopia.*
S. Simons, *CABI Africa, UN Avenue, ICRAF Complex, PO Box 633-00621, Nairobi, Kenya. Present address: Global Invasive Species Programme (GISP), UN Avenue, PO Box 633-00621, Nairobi, Kenya. E-mail: s.simons@gisp.org*
D. Teferi, *Jimma Agricultural Research Center, Ethiopian Institute of Agricultural Research, PO Box 192, Jimma, Ethiopia.*
J.M. Teri, *Tanzania Coffee Research Institute (TaCRI), Lyamungu, PO Box 3004, Moshi, Tanzania. E-mail: www.tacri.org*
D. Teshome, *Jimma Agricultural Research Center, Ethiopian Institute of Agricultural Research ,PO Box 192, Jimma, Ethiopia.*
P. Tshilenge Djim, *Université de Kinshasa, BP 117, Kinshasa XI, DRC.*
P. Wetala, *Coffee Research Centre, PO Box 185, Mukono, Uganda. E-mail: cori@africaonline.co.ug*

Preface

This is story of an African coffee farmer in Uganda who I interviewed at the start of the project at the end of 1999. This farmer had 1000 coffee trees on his 1-ha farm. He was earning US$1000 per annum from approximately 1000 kg of green coffee he produced from his field. He had a family of seven that included five children who ranged in age between 5 and 15 years. He did not hire any outside labour to assist with crop harvesting and processing, and he supplemented his income by intercropping his coffee field with bananas, which are stable food crop in the area.

In 1997/1998, when the war started in the Democratic Republic of Congo (DRC), there was a lot of coffee trafficking through illegal boarder crossings from DRC to Uganda. The military personnel operating in the area were bringing into Uganda ripe and unripe coffee cherries for processing and sale or export from Uganda. During this time, a disease called coffee wilt disease became evident to the coffee-growing public along the routes used by the military personnel. Before this period, the disease had been identified in very limited areas and was a minor threat to the coffee sector. After this period in the first year alone, the disease affected more than 600 trees from the farmer's field. In that year, the farmer's production yield was reduced by half, and by the second year, these 600 trees were dead and needed uprooting.

By 1999, the yield from the coffee farm had reduced from a peak of 1000 kg of green beans to 400 kg, representing a 60% decline in income for the farmer. In absolute terms, this represents a decline of income from US$1000 to US$400 per annum for the family of seven.

The Common Fund for Commodities has the farmer as its main client; therefore, we have every reason to worry when we witness such disasters. This project was designed to try to mitigate some of the problems by assisting the farmer described above through teaching the farmer different methods of controlling the spread of the coffee wilt disease and introducing to the farmer new disease-resistant varieties of coffee. The Common Fund has walked this journey with farmers in DRC, Uganda, Tanzania and Ethiopia and stands witness to the challenges faced by these farmers. I am happy to say that most farmers now know how to identify the disease in its early stages and how to stop its rapid spread to other trees in the field. If the farmers had the knowledge in the early 1990s, which they now have today, I am confident that at least 50% of the coffee crops loses could have been avoided.

Just to put the effects of the coffee wilt disease into context in terms of the value of coffee lost by the affected countries, you may wish to note that the affected countries export over US$1 billion worth of coffee every year. Thus, a 30% crop loss due to coffee wilt disease represents US$300 million lost income per year. In real terms, this amount represents 1 million children not having access to education, health care and sometimes basic food for survival. This project is a development lesson for policy makers who sometimes fail to appreciate the challenges encountered by farmers in the agricultural business. As Project Manager, I would like to sincerely thank all the coffee scientists, farmer group leaders and Extension Service staff who guided the smooth implementation of this project.

Caleb Dengu
First Project Manager
Common Fund for Commodities

Acknowledgements

I would like to thank all the contributors to this book- without their patience, dedication and hard work, this book would not have been possible. I would also like to thank our many collaborators including scientists and technicians from national and international research institutes and others who work in the coffee sector in the participating countries such as farmers and extensionists who participated in and helped with, many of the activities that are reported in the book. They have to live with the reality of CWD on a daily basis. Acknowledgement is also due to the funders of the Regional Coffee Wilt Programme (RCWP) including the Common Fund for Commodities (CFC), the EU through its INCO-DEV Programme as well as through the Coffee Research Network (CORNET) of the Association for Strengthening Agricultural Research in East and Central Africa (ASARECA) as well as the UK Department for International Development (DFID) through its Crop Protection Programme. I would particularly like to thank Mr Caleb Dengu (CFC) for his support and encouragement from the original development of the RCWP through to its conclusion and the final outputs (this volume and the final technical report). The ICO (International Coffee Organization) are also acknowledged as the Supervisory Body for the CFC component.

Lastly I would like to thank my colleagues in CABI notably Dennis Rangi, Mike Bodinham, Peter Baker, Noah Phiri and Sarah Cranney for their inputs and advice.

1 Introduction

Julie Flood
CABI, Bakeham Lane, Egham, Surrey, TW20 9TY

During the 20th century, a devastating disease (coffee wilt disease [CWD]) spread across Africa reducing yields, destroying millions of coffee trees in affected countries and costing hundreds of millions of dollars in lost earnings to farmers. The disease has also contributed to a decline in revenue for several African nations due to reduced coffee production and will be a contributory factor in any attempt at revitalization of the African coffee sector in the future. CWD is of particular significance because unlike many other coffee diseases, it kills trees and the farmer is often unable to replace his crop easily and faces major short falls in his cash income. The disease has reached epidemic proportions twice during the last century, becoming a serious constraint during the 1930s, 1940s and 1950s and was responsible for the complete failure of excelsa coffee commercially. However, following the systematic introduction of sanitation methods (uprooting and burning) plus comprehensive breeding programmes in many of the affected African countries, CWD was considered a minor problem. Nevertheless, the disease re-emerged again on robusta coffee in Central Africa with sporadic outbreaks observed during the 1970s and building to another extensive epidemic during the 1980s and 1990s. CWD has become an increasingly important production constraint on arabica in Ethiopia. In the intervening time between the two main epidemics, the African coffee sector had changed significantly. During the second half of the 20th century, there has been a shift from being predominantly plantation cultivation to being predominantly smallholder production, and consequently, management of the disease required reappraisal. The Regional Coffee Wilt Programme (RCWP) was a multi-country, multi-donor project initiated at the start of the 21st century to examine the disease critically and to identify suitable management options for smallholder farmers in order to reduce the incidence of the disease, minimize coffee losses and lessen impact on farmer income.

CWD, which is also called tracheomycosis, is a fungal disease. The causal agent, *Fusarium xylarioides* (sexual form, *Gibberella xylarioides*), is a vascular wilt pathogen. The fungus invades the coffee tree and colonizes the xylem system (water-conducting elements of the tree). External symptoms exhibited by coffee plants affected by CWD are generally similar to those affected with other vascular wilt pathogens. Colonization of the vascular system induces host responses, which disrupt water conduction, and this is manifested as wilting and desiccation of leaves followed by defoliation and dieback of affected branches. Symptoms may appear at any stage of crop growth, and the rate at which they develop varies. Once affected, death of the plant is inevitable, and in mature trees, it usually occurs between 3 and 15 months after the first appearance of symptoms. Young plants, however, may be killed within a matter of a few weeks of infection. Symptoms can be more pronounced on one part of the tree, a likely consequence of initial infection occurring on one of several main stems, but defoliation gradually extends to the entire plant. Coffee berries that would normally be green may redden as if ripening prematurely but often remain intact on shoots following defoliation. Other external symptoms include the swelling of the trunk and the appearance of vertical or spiralling cracks in the bark of mature trees. Small blackish-brown perithecia (sexual stage) of the fungus, similar in appearance to dark soil particles, may be produced in the cracks of the bark. Characteristic bluish-black staining of the wood can be observed directly beneath the bark.

The disease was first observed in 1927 in a plantation of coffee excelsa near Bangui in the Central African Republic (CAR), which was then known as Oubangui-Chari (Figueres, 1940). By 1945, the disease had destroyed most of the country's *Coffea excelsa* plantations (Saccas, 1951). Later, the disease was found on other coffee species (*Coffea canephora* and *Coffea neo-arnoldiana*) in parts of the CAR. It was similarly reported in several other countries in West and Central Africa around the 1930s and 1940s. In 1939, Steyaert observed the disease on *C. excelsa* plantations in the Democratic Republic of Congo (DRC; formerly known as Zaire and the Belgian Congo), close to the border with Sudan at Aba (Steyaert, 1948; Fraselle, 1950). Initially, it caused a few problems, but, as in CAR, it later reached epidemic proportions when it spread to robusta coffee. Around the same time (1938–1945), the disease also became established on excelsa in Cameroon, causing complete destruction of plantations in the east of that country. In 1947, it was discovered on *C. canephora* in Côte d'Ivoire, and major losses occurred there in the 1950s. More than 50% of the coffee-producing areas were destroyed in both Côte d'Ivoire and in DRC, and the kouillou line of robusta was completely wiped out in Cote d'Ivoire. By 1958, the disease was reported in Guinea, and it spread quickly to most of the coffee areas, causing coffee production to fall by nearly 50% (Chiarappa, 1969). In East Africa, in Ethiopia, symptoms similar to those of CWD were documented on *Coffea arabica* for the first time by Stewart (1957) and by Lejeune (1958).

At an international conference held in 1956, recommendations were made for a management programme to include (i) systematic sanitation over vast areas where affected coffee plants were to be uprooted and destroyed; (ii) where possible, relocation of coffee production to new locations; and (iii) replanting with resistant germplasm. Implementation of these recommendations reduced the impact of CWD, and literature produced during the

1970s and 1980s referred to this as a minor disease, of little importance for arabica and robusta coffee production.

However, in remote areas of DRC, the disease continued to be observed especially around abandoned plantations in the north-east of the country (Chapter 2, this volume). As early as the 1970s, farmers in Aketi (76 km from Isiro) had observed the disease in abandoned plantations, and during a survey conducted from 1974 to 1975, a number of Institut National pour l'étude et la Recherche Agronomique (INERA) fields around Yangambi were also reported to be affected. Throughout the 1980s, reports persisted of a wilt-like disease affecting coffee around the town of Isiro, and surveys were conducted (Chapter 2, this volume). There is a considerable distance between these areas, and this suggests that the disease was very widespread in that region during this time. In 1995, CAB International's (CABI) plant clinic laboratory received samples of diseased coffee plants from the Managing Director of Esco Zaire sprl (Mr. Philip Betts), and *F. xylarioides* was isolated. These samples had been collected from robusta coffee in Beni and Rutchuru, again indicating that the disease was widespread in the region. In 1995, Office Zaïrois du Café (OZACAF), now called Office National du Café, prepared a detailed report for the International Coffee Organisation (ICO) in which they outlined the extent of the problem and its serious effects on the economy of north-east DRC.

In March 1996, ICO facilitated contact between OZACAF and CABI because of CABI's experience with diseases of perennial crops including coffee. This was the start of my own personal involvement with CWD. I was invited by OZACAF to go to DRC and prepare an independent report on the nature and extent of the problem. I travelled to the north-east of DRC in July 1996 and conducted surveys of plantations and smallholder farms as well as interviewed farmers about the disease, which they called "coffee AIDS". Unlike many other coffee diseases where the yield can be reduced (for example, if the disease affects the berries) with CWD, once affected, the coffee trees died. One agriculturist I interviewed told me that he had observed the disease first in the 1970s on a large abandoned plantation on the road to Aketi, which confirms other reports that the disease had re-emerged as a serious problem in this part of Africa at that time – just when most scientific authorities considered it as a minor problem.

The diagnostic symptom of CWD – blue black staining below the bark of the coffee bush – was commonly seen, and the pathogen was isolated (Flood, 1996). The despair of coffee growers in this region was palpable. They were angry, and they wanted action. They wanted to know how to manage the problem. Unfortunately, the only recommendation then available was the same as had been given in the 1950s – sanitation, i.e. uprooting and burning coupled with long-term breeding programmes. Sanitation was being conducted on plantations, but the disease was spreading faster than plantation workers could undertake the work. For smallholders, removal of very old coffee bushes using hand tools was impossible; even where a plant had been dead for a year or more, its root system was virtually impossible to excavate.

I presented my report to the Executive Board of the ICO in September 1996 (Flood, 1996), and I was formally invited by Mr. William Naggaga (then Secretary of the Ugandan Coffee Development Authority) to make a similar fact-finding visit to Uganda because similar symptoms of a coffee disease had been observed in some of the districts bordering DRC and in Mukono. Several surveys had been undertaken by Ugandan scientists, and the pathogen had been identified (Chapter 3, this volume). It had also been noted in Uganda that the disease was only affecting robusta coffee; arabica coffee was not affected there, although the disease has continued to be a problem in Ethiopia on *C. arabica* (Chapter 4, this volume).

Following presentation of my report on the Ugandan situation to the ICO (Flood, 1997) and in recognizing the severity of the situation, an international workshop was held in Kampala, Uganda (July 28th–30th, 1997). A proposal for a funded programme of work to alleviate the problem was developed and submitted to several donor agencies including the Common Fund for Commodities (CFC), the European Union (EU) and the UK Department of International Development. Subsequently, the proposal had to be developed as a fully integrated programme of activities addressing different aspects of the disease and its management to facilitate funding of discrete activities by the various agencies, in line with their priorities and timelines, and to make tenable financial arrangements. The revisions resulted in the establishment of five interrelated and interdependent projects, each dealing with specific components of the disease. The collective name for these projects was the RCWP, and its overall coordination (by CABI) was funded by the CFC. Biological and socio-economic surveys were conducted in-country to ascertain the extent of the disease and its impact on livelihoods. These surveys were funded by the EU through the Association for Strengthening Agricultural Research in Eastern and Central Africa/Coffee Research Network. The RCWP also conducted research on remote sensing (funded by the EU), on durable resistance to the pathogen in Africa (funded by EU-INCO-DEV), as well examining the epidemiology and pathogen variation (funded by the UK Department of International Development–Crop Protection Programme and EU-INCO-DEV). In addition, considerable attention was made to the training of extensionists and farmers and to the dissemination of information about CWD to facilitate surveillance and to limit spread. This component of the RCWP was funded by CFC. For further details of the individual projects including details of implementing agencies, partners, sponsors and results, see the Final Technical Report of the RCWP (Phiri & Baker, 2009).

This book represents one output of the RCWP and, in addition to Phiri & Baker (2009), represents a compilation of the current knowledge of various aspects of CWD including information about the pathogen and its management. Although much has been achieved, considerable work remains. Over the 13 years since I conducted surveys in DRC, the disease has spread much more extensively within DRC (Chapter 2, this volume) and has been reported in the province of Equateur, threatening coffee production in Western Africa. It has continued to spread throughout all coffee-growing districts in Uganda (Chapter 3, this volume) and into Tanzania (Chapter 5, this volume). The disease continues to be a problem of *C. arabica* in Ethiopia (Chapter 4, this volume). Severe impacts on the incomes and livelihoods of coffee farmers in affected areas of DRC, Uganda, Ethiopia and Tanzania have been observed, and producers have had to diversify away from coffee production by either growing other crops or engaging in non-agricultural activities (Chapter 6, this volume). Early observations suggested that in areas where both coffee species existed, only one species was affected, and artificial inoculation has confirmed that there is host specificity of the strains from *C. arabica* and *C. canephora* (Chapter 8, this volume). In addition, strains from arabica isolated from Ethiopia are pathogenic to *C. arabica* germplasm from other coffee-producing areas (Chapter 8, this volume). These two distinct populations were also discernable using various molecular methods, although the overall variability appears to be low despite the production of perithecia in nature (Chapter 7, this volume). As part of the RCWP, participatory trials on managing the disease were undertaken in many locations throughout affected countries (Chapter 9, this volume) so that farmers could search for ways of reducing the impact of CWD. The effect of wounding coffee trees on increasing the incidence of the disease was quickly appreciated. These participatory trials were also part of a need to raise awareness of the disease across the region so producers and others, such as extensionists and researchers, are able to quickly recognize its symptoms. If the disease spread to new areas, then for eradication to have any chance of being effective, early detection is critical. Thus, the RCWP undertook extensive training of farmers and of extensionists (as trainers) in participating countries using the Farmer Field School approach, which was adapted for coffee (Chapter 11, this volume). The Farmer Field School approach plus field days and workshops proved to be effective tools for facilitating farmers' learning and in enhancing their critical thinking and problem-solving skills. Mass communication methods such as print materials and use of radio and television were also used and discussed (Chapter 11, this volume). Farmers in affected countries are now much more familiar with the disease and its management. Breeding for resistance remains the best option for management of this disease (and was successful in earlier epidemics in the 20th century). Programmes for screening for resistance have been initiated in participating countries, and resistant material has been identified. In Uganda, several resistant genotypes have been selected and are currently being multiplied for distribution to farmers (Chapter 10, this volume).

References

Chiarappa, L. (1969) International assistance in plant pathology in developing countries. With particular reference to FAO programmes. *FAO Plant Protection Bulletin* 17 7–8.

Figueres, R. (1940) Sur une maladie très grave du caféier en Oubangui. Rapport. Ministère des Colonies, Paris, France.

Flood, J. (1996) A study of tracheomycosis or vascular wilt disease of coffee in Zaire. Report presented to Zairean Coffee Organization (OZACAF). August 1996. 13 pp.

Flood, J. (1997) Tracheomycosis or vascular wilt disease of coffee in Uganda. Report presented to Ugandan Coffee Development Authority (UCDA), Uganda. 12 pp.

Fraselle, J. (1950) Observations préliminaires sur une trachéomycose de *Coffea robusta. Bulletin Agricole du Congo Belge* XLI, 361–372.

Lejeune, J.B.H. (1958) Rapport au Gouvernement Imperial d'Ethiopie sur la production caféière. Rapport du la FAO, Rome FAO158/3/1881.

Phiri, N. and Baker, P.S. (2009) Coffee wilt disease in Africa; a synthesis of the work for the Regional Coffee Wilt Programme (RCWP). Final Technical Report. CABI. 200 pp.

Saccas, A.M. (1951) La trachéomycose (carbunculariose) des *Coffea excelsa, neo-arnoldiana* et robusta en Oubangui-Chari. *Agronomie Tropicale* 6, 453–506.

Stewart, R.B. (1957) Some diseases occurring in Kaffa Province, Ethiopia. In: *Imperial Ethiopian College of Agriculture and Mechanical Arts.* Alemaya, Ethiopia, pp. 15–16.

Steyaert, R.L. (1948) Contribution à l'étude des parasites des végétaux du Congo Belge. *Bulletin de la Societe Royale de Botanique de Belgique* 80, 11–58.

2 Coffee Wilt Disease in Democratic Republic of Congo

A. Kalonji-Mbuyi,[1,2] P. Tshilenge Djim[1] and N.T. Saiba[3]
[1]*Université de Kinshasa, BP117, Kinshasa XI, Democratic Republic of the Congo*
[2]*Centre Régional d'Etudes Nucléaires de Kinshasa (CREN-K), BP 868, Kinshasa XI, Democratic Republic of the Congo*
e-mail: adrienkalonji@yahoo.fr
[3]*Office National du Café, BP 8931, Kinshasa, Democratic Republic of the Congo*

2.1. Introduction

Coffee is one of the key cash crops in the Democratic Republic of Congo (DRC). Several *Coffea* species are found across the country. The first coffee plantations grown in DRC were from spontaneous species identified in some provinces. The main exotic species include *Coffea abeokuta*, *Coffea excelsa*, *Coffea myrtifolia*, *Coffea stenophylla* and some varieties of *Coffea arabica* imported from Belgian Congo (now DRC) by the colonial garden (Kinds, 1930). *Coffea liberica* from Liberia was the first introduced *Coffea* species in DRC in 1881 (Jagoret and Decroix, 2002). However, commercial coffee production relies mainly on two species, *C. arabica* L. (70%) and *Coffea canephora* Pierre (30%). The main development of the coffee crop in DRC took place between 1924 and 1930. This was made possible because of an important collection of local *C. canephora* accessions assembled mostly from prospection in the forests of the Yangambi and Lula regions, as well as in other parts of Congolese and foreign zones.

Discovered in 1898 in what is now DRC, *Coffea robusta* is the most commonly grown variety of *C. canephora*, accounting for up to 95% of *C. canephora* populations worldwide (Miny, 1930). Unlike *C. arabica* plants, *C. robusta* does not need to grow at high altitude and requires less care to grow because it is hardier and tends to be less susceptible to pests. Its area of distribution is variable and corresponds to hot and humid climatic regions. It is found in low- and middle-altitude areas in Africa (Ivory Coast, DRC, Cameroon, Uganda, Angola, Ghana, Togo, Madagascar, Tanzania and Republic of

Central Africa), in the Far East (India, Indonesia, Philippines, etc.) and in Oceania, for example, New Caledonia.

Most of the robusta coffee plantations in DRC derive from progenies of a few elite local clones from the Yangambi Research Stations, and one introduced clone (SA 34) from Indonesia, that were previously selected for yield. No specific breeding programmes were developed to improve the agronomic traits of genotypes, but coffee selection based on yields and morphological traits was carried on by the 'Institut National d'Etudes et de Recherches Agronomiques' (INERA). Six elite local clones including L-147, L-36, L-251, L-215, L-93 and L-48 were selected from monoclonal isolate plots (Drachoussof *et al.*, 1956). This standard mixing continues to be highly productive and provides financial sustainability to coffee producers.

Robusta coffee is of paramount economic importance in DRC. It represents almost 90% of all the plantations across the country and is DRC's biggest cash crop. It is with maize and rice as the top leading crops. Robusta coffee is the third exported product in DRC after copper and cobalt, and it generates nearly 9% of the total revenue from exportation for the country. This represents almost 60% to 65% of the total value of exported agricultural products (Banque Centrale du Congo, 2005). Coffee is produced at village and agro-industrial levels (Petites et Moyennes Entreprises Agricoles). At village level, coffee generates for local producers considerable cash flow estimated at 2 billion Congolese franks (Office National du Café [ONC]). This represents on average nearly half of the income generated from coffee exportation zones (National Agricultural Statistic Service, 2006, unpublished report).

2.2. Importance of Coffee in DRC

Coffee is an important commodity in the economy of DRC. The coffee species grown in DRC are *C. arabica*, *C. canephora* var. *robusta* and *C. canephora* var. *kouillou*. The spatial distribution of coffee farming is dependent on ecological conditions that are favourable for their cultivation. The current total area under coffee covers nearly 193,433 ha. The estimated area under robusta, including the Kouillou, is about 164,533 ha (85. 1%), and arabica coffee covers 28,900 ha (14.9%). Most coffee farmers are smallholders, which represent around 86% of the total production with average fields per farmer of 1.3 ha for robusta and 0.8 ha for arabica. From the agricultural census, it has been estimated that a third of the coffee plantations are now abandoned.

Production of coffee has been fluctuating for many years. For the 2006–2007 season, production was estimated at 34,553 t, with 27,007 t for robusta and 7546 t for arabica (ONC, 2007, unpublished data). The national average coffee production varies between 150 and 200 kg/ha for robusta and between 150 and 300 kg/ha for arabica. This is produced on 193,433 ha of land, so the productivity is low. At the time of its independence in 1960, the country was the leading African coffee exporter, with a production of 52,000 t of coffee annually. Coffee production subsequently increased, reaching a

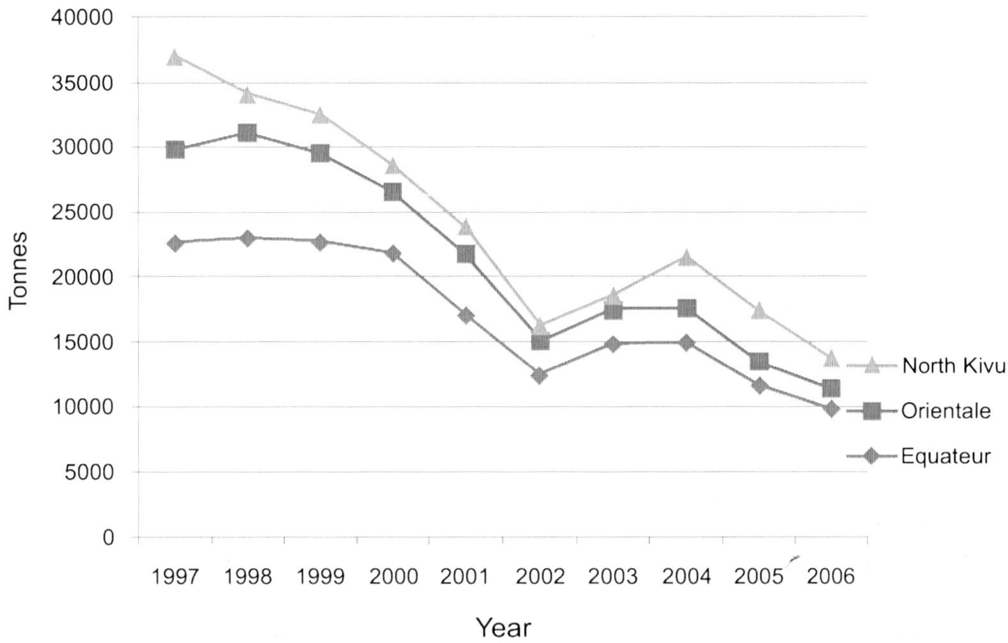

Fig. 2.1. Annual coffee production (var. *robusta*) in the infected provinces by CWD in DRC.

record level of 130,000 t from 1989 to 1990, comprising 25,600 t of arabica and 104,400 t of robusta, including Kouillou (Conjoncture Economique, 1991; ONC, 1990, unpublished data).

The data of the average robusta coffee production in Equateur, North Kivu and Orientale province areas for the period 1997–2006 are summarized in Fig. 2.1.

Coffee production decreased over the last decade, although there was some increase in production in 2003 and 2004. Production then decreased again rapidly, presumably due to the continuing impact of the coffee wilt disease (CWD).

2.3. Constraints to Coffee Production

The coffee sector in DRC faces a number of challenges of diverse nature. Notable challenges include the degeneration of the existing planting material, the aging of the coffee gardens, the poor management of plantations, the lack of skills and knowledge of coffee production by smallholder producers and an inadequate, poor and dilapidated road and port infrastructure. In addition to these conditions, the potential of the sector has been progressively reduced due to the following: the effects of nationalization of the coffee production sector in 1974; the continuing low farm gate price of coffee, which is a worldwide phenomenon; the drift of people from the land into the cities; and for those that remain on the land and try to farm, they are faced by a lack of credit.

2.4. Most Important Diseases and Pests

Although socioeconomic and political conditions have changed, environmental conditions (climate, soil, etc.) remain very favourable for coffee production. However, the absence of any plant health surveys as a surveillance mechanism for alerting the authorities to the presence and subsequent upsurges in plant diseases and the absence of management methods have favoured the resurgence of numerous diseases in the Congolese coffee culture; some of which have reached epidemic proportions. Among these diseases is CWD or tracheomycosis caused by *Gibberella* (*Fusarium*) *xylarioides* Heim & Saccas. When surveys were subsequently undertaken, the presence of anthracnose (*Colletotrichum* sp.) and a root rot caused by *Armillaria mellea* mainly in Bas-Congo province in the Bas-Fleuve district (A. Kalonji-Mbuyi and P.M.L. Onyembe, 1996, unpublished data) were also detected. The anthracnose induced large brown spots on the leaves with or sometimes without a yellow halo. Leaves fall, and the young branches wither, causing the whole plant to dieback. In some instances, the berries are also attacked. This defoliation was observed at all stages, even at the nursery. The root rot was frequently observed in the Seke-Banza zone in the Bas-Congo province. The symptoms are essentially the sudden and general wilting of the aerial part of the plant immediately followed by plant desiccation. This disease presents very similar external symptoms to those associated with CWD. The diagnostic root rot symptom is the presence of splits on the lateral roots that show the fungal mycelial net characteristic of *Armillaria*. When the root rot becomes very severe, the plant falls down if shaken or pushed.

2.5. A Brief History of CWD in DRC

CWD was first observed in 1939 by Steyaert on *C. excelsa* materials collected in plantations located at the borders between DRC and Sudan (most precisely at Aba) and others from Bangui in the Central African Republic (Fraselle, 1950).

In the beginning, the disease was minor with limited economic impact. However, around 1949, an epidemic was reported on thousands of robusta-type plants in Yangambi plantations (Heim and Saccas, 1950). The disease was later observed in Haut-Uélé, North Kasai and Katanga. Its massive reappearance during the following year illustrated its danger not only because of the considerable damage it causes but also because of its epidemic character (Fraselle, 1950). Initial CWD symptoms seen included a generalized chlorosis of the leaves, which become flaccid and curl inwards, but on other occasions, the leaves do not become yellow but wilted while remaining green. The infected tree quickly becomes defoliated. Prior to their fall, the leaves often darken, especially those on the seedlings and the nursery stage.

A number of measures to contain the disease were initiated in Yangambi City and Yangambi station, including uprooting and burning of infected coffee trees on the spot. These practices were followed, and the disease was successfully controlled. CWD was subsequently forgotten and lost its economic importance because of the development of resistant robusta varieties (Fraselle and Geortay, 1952). Successful adoption of sanitation practices and resistant varieties led to the decline of incidence of CWD from the late 1950s.

In the late 1970s, a farmer in the north-east of DRC observed a wilt-like disease of coffee in abandoned plantations around Aketi about 76 km from Isiro (Pochet, 1986, personal communication). Later, the disease was observed on coffee plants during a survey conducted from 1974 to 1975 in a number of INERA fields at Yangambi (A. Kalonji-Mbuyi, 1975, unpublished data). Studies of the samples collected by Kalonji-Mbuyi recovered a strain of *F. xylarioides* from infected plant branches, stems and roots. The identity of the fungus was later confirmed by Professor Felix Pierre-Louis to be *F. xylarioides*. Because the disease was not observed in the surrounding farmer's plantations, it has been supposed that it was due to the heterogeneity of the material used by farmers, some of which may be resistant to this fungus, while INERA's genetically uniform and homogenous plantations were considered to be more likely to be affected.

Throughout the years 1980 to 1981, national reports indicated the presence of some source of CWD around the town of Isiro in the region of Haut-Uélé, Orientale province (ONC, 1982, unpublished report). These reports indicated that the disease attacked only robusta coffee plantations and that both young and old established plantations were affected. In addition, the disease occurred both in plantations planted on rich soil and in localities with less-fertile soils.

The disease became widespread in all the Haut-Uélé plantations, progressively spreading to all the surrounding territories. More plantations became abandoned, and smallholders became desperate as they saw their only source of income disappear. The existence of a very marked and decreasing gradient of infection as one moved away from the triangle formed by Isiro territories, Wamba and Mungbere in the valley of Nepoko led to the conclusion that this area is the primary source of the infection.

Between 1985 and 1989, ONC initiated surveys at Ituri, Bas Uelé and Haut-Uélé districts with the objective of obtaining information on the disease and of the environmental factors that influence the disease. Disease surveys revealed that the disease was also widespread in Orientale province. According to estimates determined in 1987, its incidence was 19.3% of these coffee areas (M.K. Katenga, 1987, unpublished data). Two years later (1989), it reached 30.6% mainly in the district of Haut-Uélé (A. Kalonji-Mbuyi, K. Mukuna, R. Masozera and N. Isungu, 1990, unpublished data).

The progression of CWD from Haut-Uélé has been very irregular. The disease continued to spread down through a corridor of robusta coffee, which stretches from Mambassa to Irumu in the Ituri district (Fig. 2.2). The incidence of infection was reported then as being around 30% in 1992

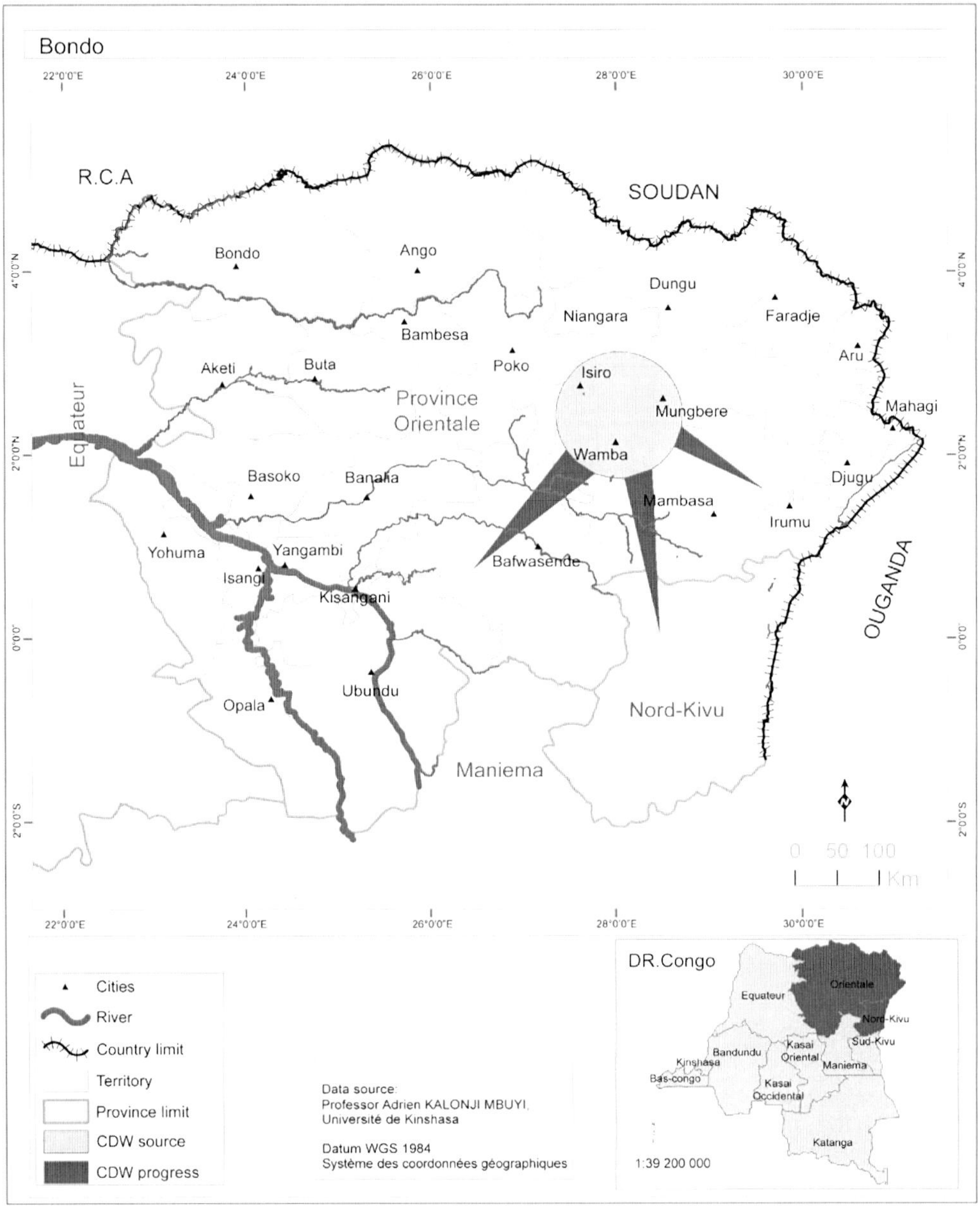

Fig. 2.2. Primary source of CWD Infection in DRC (Isiro-Wamba-Mungbere) and progress to North Kivu.

(Kalonji-Mbuyi *et al.*, 1990; Pronalutra, 1992, unpublished report). The disease then spread towards the North Kivu province, and over the following 3 to 4 years, it was consistently observed in that area from 1995; its appearance was always accompanied with considerable damage. The incidence of infection in 1995 was estimated to range from 13% to 30% of plants in plantations located in Mavivi, Mbawu and Mutwanga. However, the disease remained less serious in the surrounding areas of Butembo-Muhangi and Musienene in the Lubero district of Lubero and did not exceed 3%.

In 1996, ONC, then known as OZACAF, invited Dr Julie Flood (CABI) to assess the epidemical situation in eastern DRC, and she confirmed the presence of CWD caused by *F. xylarioides* (Flood, 1996). In 1997, assessments indicated that the incidence exceeded 50% in the coffee in the Haut-Uélé district (Tshilenge-Djim *et al.*, 1998).

Currently, CWD is a serious constraint that has caused considerable damage to coffee production in DRC, and its effects compromise the very existence of coffee growers. The disease is very widespread in the country, and its spectacular progress has significantly disrupted production levels of robusta. In its epidemic phase, the disease has manifested itself as destroying many of the coffee plantations in the greater production zone of DRC. In certain places, it appears to be endemic with little infection sources, too scattered and difficult to detect by the planters. Nevertheless, these small groups (disease foci) may represent a future threat to any future coffee material.

2.6. CWD Distribution in DRC

In DRC, CWD attacks robusta coffee and other coffee species such as *C. liberica* (*C. liberica* var. *liberica*, *C. liberica* var *abeokutae*, *C. liberica* var. *dewverii*) in the Botanical Garden collections at Yangambi station (INERA). Several visits to the arabica coffee area showed that this species is not affected.

A biological survey was carried out between 1991 and 2004 in different plantations across the seven coffee-producing provinces in DRC: Bas-Congo, Bandundu, Equateur, Kasai Occidental, Kasai Oriental, Nord-Kivu and Orientale province. The general and specific objective of the survey was to determine disease incidence on the coffee trees in the plantations within the production area. The survey involved controlled sampling of coffee plantations to collect information on plant health, state of the coffee trees, the presence or absence of CWD, and in the case of the disease being present, to determine the incidence, severity and relationships with the ecoclimatic and environmental conditions so as to contribute to the design of control strategies adapted to the various conditions. The methodology used to collect data was based on the identification of infected plants in the plantations by taking into account the external symptom expression (yellowing, drying of the crown) and the internal characteristics (presence of the blue–black bands on the level of the collar of the plant or in the aerial parts). Principal sampling of the inspected plantations was undertaken from the plantations chosen randomly in groupings, traditional entities gathering several villages.

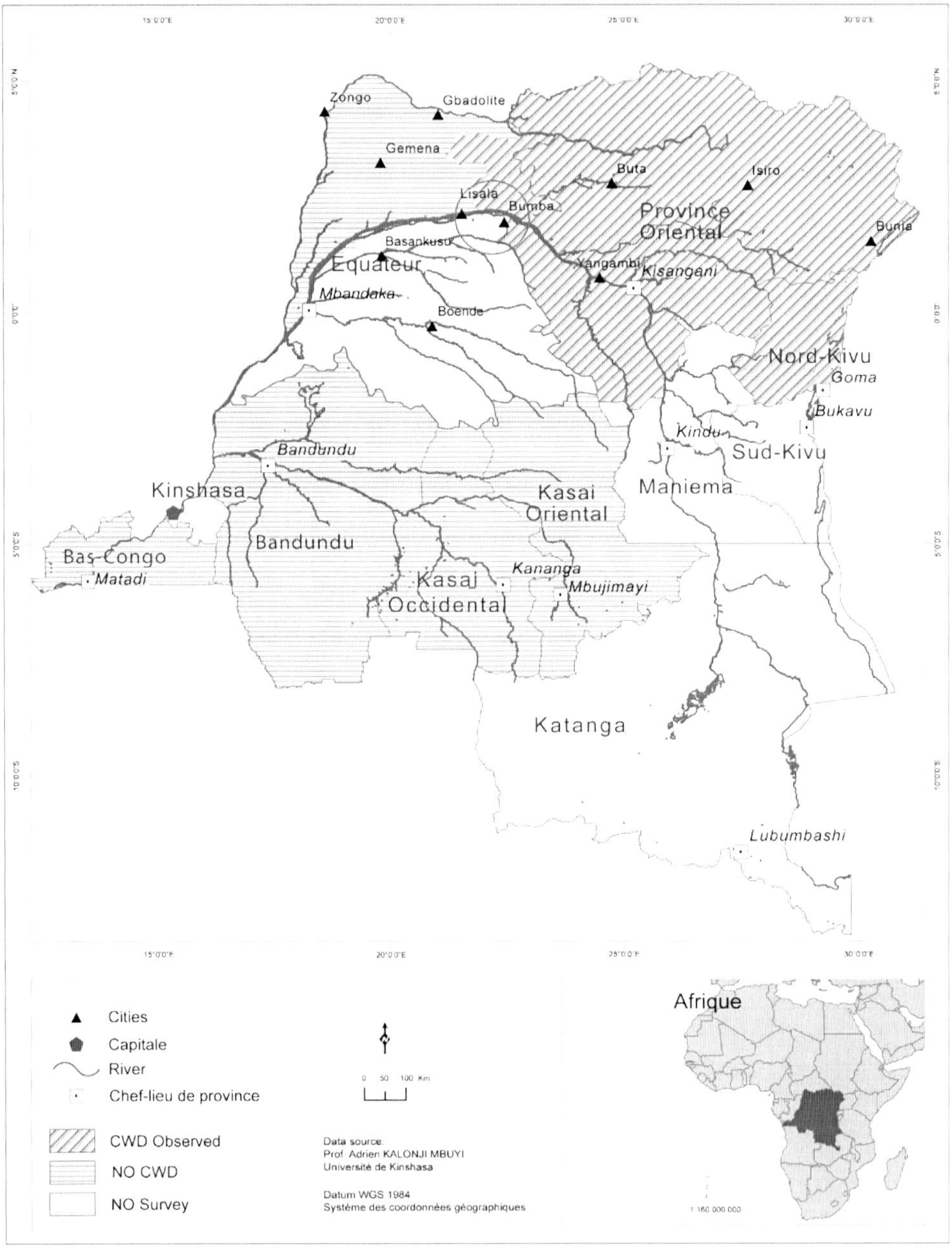

Fig. 2.3. The coffee wilt disease distribution in DRC.

Large commercial plantations were rare, and those still existing were subdivided into several groups, in which three to six were chosen at random for surveying.

The ONC coordinated the survey, and scientists who took part included plant pathologists from the University of Kinshasa and from the INERA; staff from CABI Africa were the enumerators. The team members visited the farms selected, with farmers doing the data collection. In each farm visited, 30 trees in a randomly selected diagonal or median direction were scored. Biophysical factors such as production system, age of the coffee, altitude, topography and information about the cultural practices (shade, fertilizer use, intercropping, mulching, etc.) were also collected to assess their possible influence on CWD incidence and severity.

The surveys revealed that CWD is present in North Kivu, Orientale and Equateur provinces (Fig. 2.3). The disease was not observed in the provinces of Bandundu, Bas Congo, Kasai Occidental, Kasai Oriental and South Kivu.

2.7. Importance of CWD in DRC

Results obtained in the infected zones for the periods 1991–1996 and 2002–2006 showed that a lack of an adequate plant health service is a major contributor to the establishment of CWD in epidemic proportions in these areas. CWD has increased its frequency, forming epidemic levels because the symptoms on a few isolated diseased plants were not recognized rapidly enough, and even when diagnosed, the management techniques were not implemented quickly enough.

The data obtained (Fig. 2.4) varied from site to site, although similar rates of infection were observed. For all the observations throughout the period 1991–1996, the highest values in all the sites were registered in 1996.

In Orientale province, Isiro had 90% of the plantations affected, Mambassa had 36%, Poko had 34%, Bafwasende had 28%, Opala had 29% and Banalia had 27%. In this period, the Yangambi station was disease-free. For the same period, in the province of North Kivu, the disease incidence varied according to sites, with 46% of the plantations in Mangina affected, 41% at Oïcha, 38% at Muhangi and 37% at Mutwanga.

Two distinct periods are observed when considering CWD evolution. The evolution of the disease between 1994 and 2002 has followed the normal epidemic disease progress expanding to reach the maximum values for all the investigated sites. The observations carried out in 2004 revealed a revival of the infection cycle for all sites, and this revival was, in fact, faster, for example, 73% infection was observed at Oïcha. Such type of epidemiologic evolution confirms the polycyclic feature of tracheomycosis on coffee in DRC.

When ground surveys could not be conducted due to periods of political instability (1996–2006), a simulation of the disease progression was made by applying logistical models so as to forecast the disease situation. Comparisons between the disease incidence forecasted with the actual observed values in 1995, 2002 and 2004 indicated some deviations in the actual

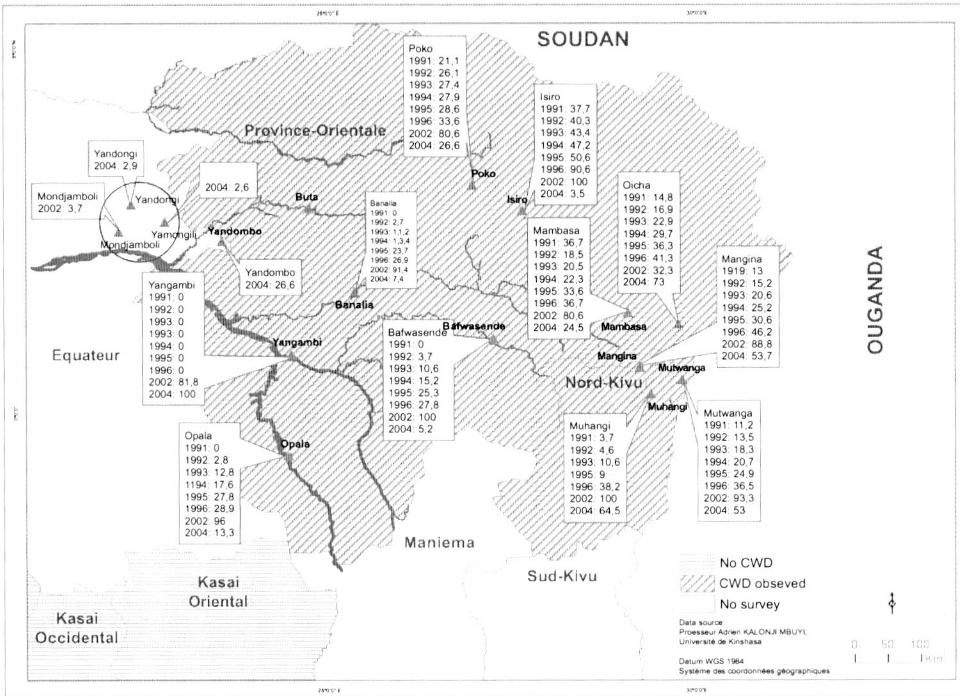

Fig. 2.4. Evolution of CWD in the infected zone between 1991–2004

observations compared with the forecasts at some sites. Those deviations are due to various reasons according to the sites. Sites such as Bafwasende, Opala and Banalia (Fig. 2.5a–d), which are located in Central Basin in the Orientale province, were characterized by low altitude (less than 400 m) and heavy annual rains (superior to 188 mm), and there, the observed and simulated evolution of CWD are similar. However, in Mambassa, the logistical model did not accurately predict the situation of sigmoid development of the CWD epidemic when we consider data for the 1995–1996 period. This was probably because of plantation abandonment.

At Isiro (Fig. 2.5e), the development of CWD shows a difference between the values observed in 1996 and the forecast, with the model reproducing a disease level for the site, which was already observed around 1983 (M.K. Katenga, 1987, unpublished data).

In the province of North Kivu (Fig. 2.6a–d), there are two situations. At Muhangi, the development of CWD was very different between observed and simulated values. It is worth noting for this site that the farmer gave up maintaining plantations of robusta coffee in 1996 and adopted arabica coffee. The Muhangi site is located in a high altitude of more than 1300 m, which is favourable for arabica culture. The development of CWD, as revealed by the model, suggested that with routine maintenance practices and uprooting of diseased plants, the source of inoculum for the subsequent infection would have been around 10.5% in 1996 (according to the model), but 38.2% infection

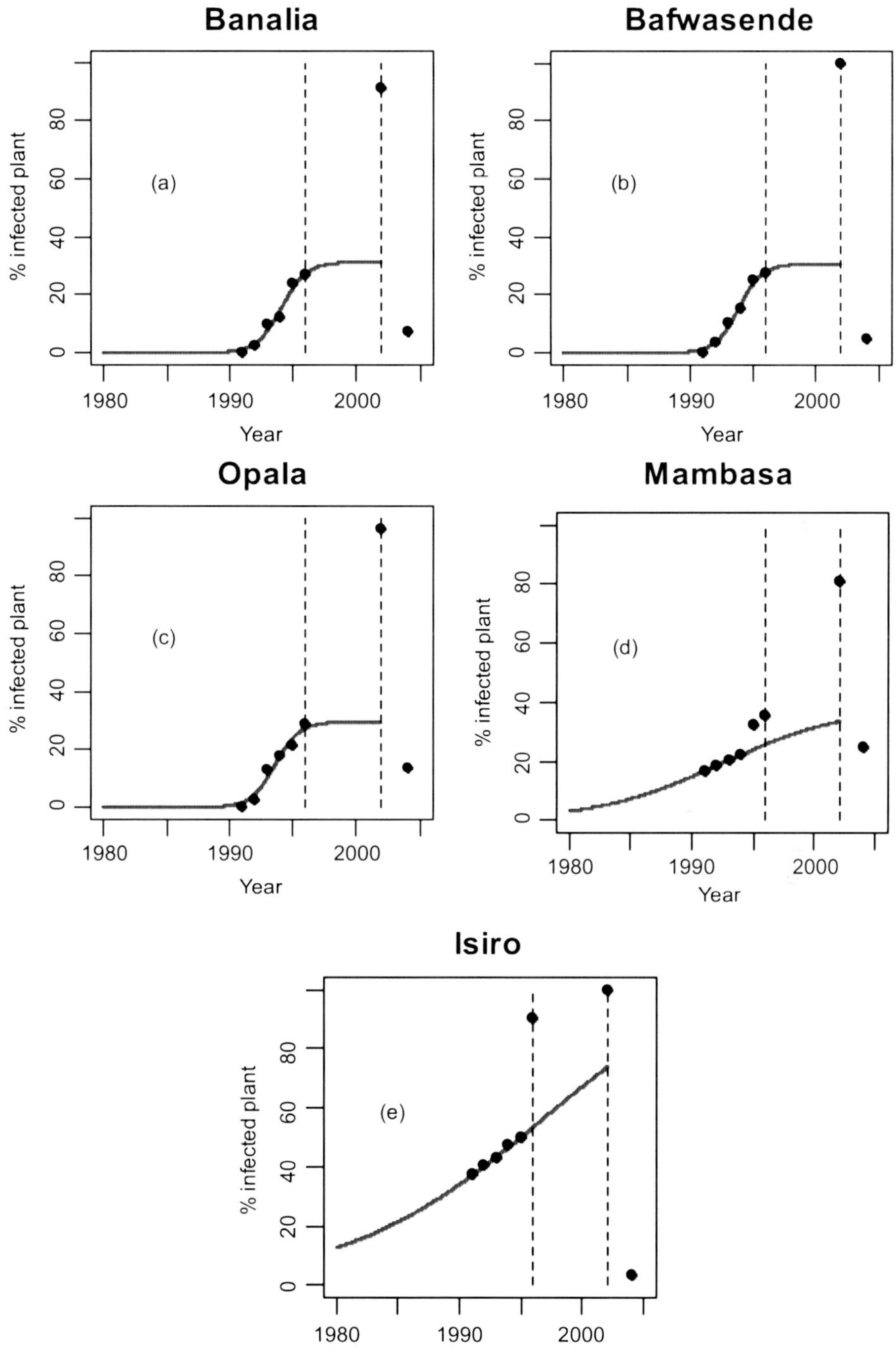

Fig. 2.5. Evolution of CWD in DRC by applying logistical models: (a) Banalia, (b) Bafwasende, (c) Opala, (d) Mambada and (e) Isiro.

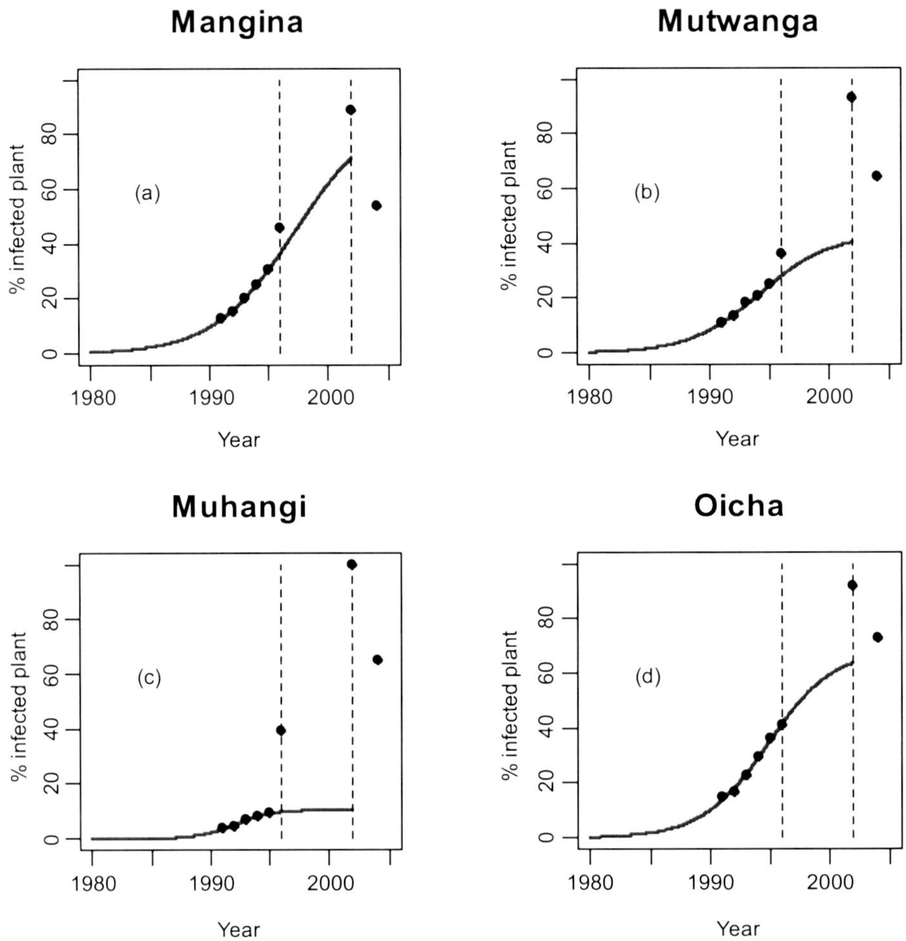

Fig. 2.6. Evolution of CWD in DRC by applying logistical models: (a) Mangina, (b) Mutwanga, (c) Muhangi and (d) Oicha.

was actually observed. The sites of Oïcha, Mutwanga and Mangina demonstrated similarities between the model and observed values with slight gaps in data for the period 1995–1996. Any large deviations are attributable to the abandonment of plantations due to civil wars in those regions.

From the simulation studies, it seems that if all the factors required to stop the development of CWD were maintained in all the sites throughout the period (1991–1996), then the development of the disease would have reached maximal level in the years 1995 to 1996. However, in reality, in most of the cases, after 1996, CWD reached a peak in north-east DRC, which corresponded to its maximal level, and a new disease cycle started on young replantings in 2004.

The abandonment of almost all the plantations due to civil wars that have occurred in that part of the republic seems to have favoured the development of the disease. The case of Yangambi might serve as an example, where CWD was almost absent until 1996 and its sudden reappearance in 2002 with an increased severity (81.8%) and incidence (100%).

Presently, CWD has infected plantations in the province of Equateur through the district of Mongala on the border with Orientale province, and it has progressed into the province to where the latest observations (Kalonji, 2007, unpublished data) indicate its presence in Gemena, a town located in the north-west.

2.8. Factors Affecting Severity of CWD in DRC

There are many factors that influence the incidence and severity of CWD in DRC.

2.8.1. Agronomic and environmental factors

Analysis of some agronomical and environmental conditions has shown that CWD is a disease that manifests itself in many diversified environments including valleys, forest zones and high and low locations. The effects of human influence through agronomic practices vary from province to province. The age of plantation varies between 5 and 15 years in North Kivu (88%), whereas in Orientale and Equateur provinces, most of the plantations are 15 years old (75% and 85%, respectively). Age is a factor connected to the presence of CWD in the three provinces where this disease exists. The presence of CWD on all plantations of over 15 years in Bas-Congo (81.6%) and in Bandundu (64.6%) shows that age could be a major risk factor, and the probability of infection is increased each time the coffee trees are pruned. Fraselle and Geortay (1952) reported that the period of the first production would be a favourable circumstance for the extensive development of the disease, although it can also affect the young plantlets in the nursery.

The advanced age of the plantations necessitates replanting to improve the productivity. Most farmers use locally produced seedlings to plant new fields. Ninety percent of farmers in Bas-Congo, 80% in Bandundu, 82% in Kasai Occidental and 53% in Kasai Oriental use this practice. In North Kivu and Orientale province, the local germplasm is the most frequently used (89.8% and 65%, respectively). Generally, the use of seedling can have an important influence on the development of CWD. Seedlings normally have a high level of heterogeneity, and such diverse germplasm could include those genotypes that have some resistance to the disease. However, the local material used in the affected provinces could have a restricted genetic base due to selections being constantly taken from the farmers' own fields or plantations.

Furthermore, in the three provinces where CWD is present, the production system is mainly of the large-scale plantation system (78.5% in Orientale province and 59.5% in North Kivu) followed by the semiplantation system (54.6% in Equateur province). These two production systems favour the disease.

Weed management is another factor affecting CWD; most of the plantations do not have weed management (81.4% in Equateur and 44% in Orientale). This could have an inverse relationship with the development of the disease, because in the provinces where CWD is absent, such as Kasai Oriental, Kasai Occidental and Bas-Congo, this agronomic practice is very frequent (54%, 44.3% and 43.6%, respectively).

With regard to the influence of pruning on the incidence of CWD, the results are contradictory. For instance, in Orientale province, pruning, which is practiced by 80% of the farmers, would have favoured the spread of the disease, whereas in North Kivu, where 84% of the farmers do not practice pruning, the incidence of CWD is higher (mean of 90% and 94% of infection, respectively, for Orientale province and North Kivu). Similar results have been reported by Fraselle and Geortay (1952), who indicated that the pruning practice does not appear to significantly influence disease incidence. The authors however have observed that at Yangambi, where coffee bushes have two or three stems (multicauline pruning system), an accelerated spread of CWD was observed. This might be due to the many wounds created by this pruning system, which enhance the pathogen penetration of the coffee tree. The practice can be safe if applied using sharp and disinfected tools. Other agronomic practices such as shade, fertilizer use and mulching were similar in all the surveyed regions, so any differential effects on CWD were not apparent.

It is not obvious to find a link between intercropping and CWD. Intercropping is more frequent in North Kivu (71.4%), in Kasai Occidental (53.3%) and in Orientale province (52.5%), whereas it is less frequent in Kasai Oriental (27.8%), in Bas-Congo (20.4%) and in Bandundu (12.9%). Other factors such as environmental conditions (valley, forest zones, hilltop and flat landscape) do not seem to have a particular influence on CWD because of the presence of the disease in these various conditions.

2.9. Country-Specific Management Practices in DRC

2.9.1. Cultural practices and chemical control

The main strategy used in the management of CWD in DRC involved the systematic elimination of all the infected coffee trees once the most typical external symptoms of CWD (yellowing, browning unilateral or general of young shoots, fading and wilting) are observed. If conducted judiciously, then the method has the advantage of reducing pathogen inoculum at the primary centre of infection, thus preventing the spread to neighbouring healthy trees. For the method to be effective against CWD, the following observations and operations are recommended (Fraselle *et al.*, 1953):

- For each field, a detailed plan locating each coffee tree is required.
- Regular monitoring is required to detect any coffee bush showing even partial dieback symptoms on the aerial parts of the plant; all doubtful

plants are clearly identified using visible signs (string, color material or any other marking).
- After identifying a plant suspected to be infected by CWD, it is advisable to work in group to confirm whether or not that coffee tree is actually infected.
- The disease can be confirmed if the superficial bark is removed on all sides of the stem, and blue–black bands are seen below the bark; this is diagnostic of CWD.
- If the suspect tree does not have these characteristic blue–black staining below the bark, then the wounds need to be disinfected and the knife used treated with a chemical or heat treated.
- If the suspect tree is infected with CWD, then the plant needs to be clearly marked to identify its location. A spray of 10% carbolineum preparation in water solution (about 2 l of solution) should be applied to the aerial parts of the infected coffee bush.
- The infected plants should then be dug out (including the root system). The tree should be cut into sections and incinerated *in situ*. If the wood is dry, then this can be done immediately, but if the wood is still green, then a few days may be needed to allow the wood to dry to facilitate burning.

Trials undertaken *in situ* where the plantations are characterized by the absence of CWD symptoms, but close to infected plantations, allowed an assessment of the impact of some agronomic practices and management methods on both the expression and the development of the disease. Protected plants were left to natural contamination, and the assessments consisted of recording the numbers of infected plants on a monthly basis.

The agronomic practices tested consisted of clean weeding, slashing and pruning. The clean weeding was obtained by spraying herbicides glyphosate at 0.1%. The spraying was done in such a way that the herbicide did not touch the coffee tree, that is, by keeping a distance of about 30 cm from the tree. Slashing, which is a common practice of weeding, was made by using the machete. Pruning by means of shears or saw was undertaken. Chemical control was undertaken by monthly spraying of Cupravit (copper oxychloride), a fungicide, at 0.5% (wt/vol) on the base of the coffee trunk and by painting the stem with the same fungicide (0.3% [wt/vol] solution) applied to the trunk at a height of 0.5 m from the soil.

Monthly spraying of fungicide was more effective in reducing the attack of CWD than other treatments (Fig. 2.7). This was followed by slashing and clean weeding, which also seem to significantly reduce CWD incidence. Treatment using stem painting and pruning did not seem to be so effective. The high prevalence of infected trees, scored in the case of the plots under pruning (6%) and the control plots (7%), demonstrates the danger that might be associated with wounding coffee trees during routine maintenance, and the lack of maintenance or phytosanitary practices in the abandoned plantations is also a key cause of the rapid spread of CWD.

The foregoing results obtained in DRC show that besides fungicide spraying, agronomic practices that are aimed at maintaining a clean plantation

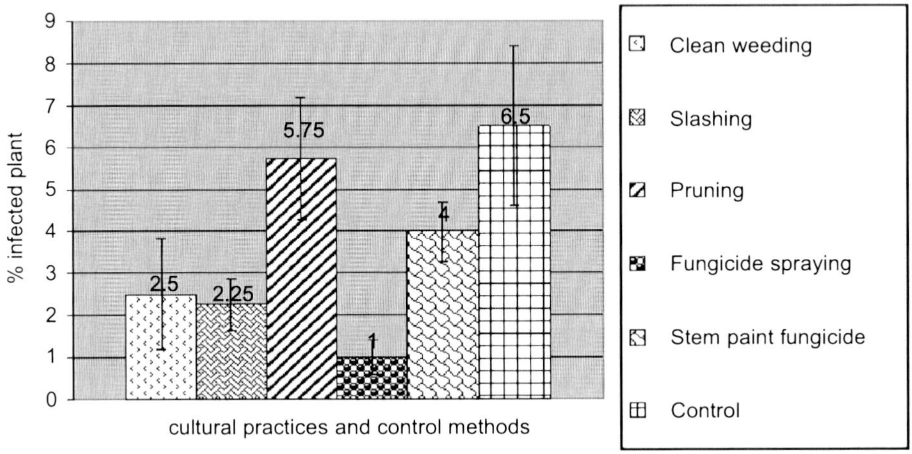

Fig. 2.7. Incidence of CWD in the robusta coffee recorded after 18 months of application of different cultural practices and control methods.

such as clean weeding and slashing would, if conducted efficiently, significantly contribute to reducing CWD spread. Alternatively, any practice that involves wounding, such as pruning, is likely to increase the rate of spread of the disease. Phiri (2005, personal communication) reported the results registered in Ethiopia, Uganda and Tanzania, where the same study was conducted at the same period as in DRC, and concluded that the prevention of any wound contributes to reducing and slowing down the spread of CWD in plantations. However, the use of chemical treatment cannot be recommended for our farmers' fields, given the cost and the danger that it represents for humans and the environment. Agronomic practices such as slashing, if conducted appropriately, for example, by avoiding wounds on coffee trees or ensuring that plantations are well maintained, will remain the main method for the management of CWD in DRC and can be easily adopted by farmers.

2.9.2. Availability of the improved material for replanting

The current situation in DRC with respect to CWD requires rehabilitation of existing production areas and the extension of these production areas using planting material of desirable characteristics. Currently, in DRC, when starting new plantations, farmers seldom use seeds from the research centres that have guaranteed characteristics for good production. Generally, the vegetative material used is from cuttings, and often, this material performs poorly. However, the elite clones selected at Yangambi, which include seven clones of robusta (L 36, L 48, L 93, L 147, L 215, L 251, SA 158) and five clones of small Kouillou (LAF 93, LAF 159, S 9, S 19, S 23), are available, and those can be used preferentially by the farmers.

One of the priorities of the DRC government, as described in the Strategic Action Plan, is the revival of the coffee production by the systematic replanting of the old plantations. The government proposes to rehabilitate all the centres of production and the multiplication of the elite clones of INERA. The plan also includes the reorganization of the supply chains of quality coffee material by the creation of multiplication centres and the dissemination of improved material in each of the four principal zones selected above. This will require the involvement of the Ministry of Agriculture (technical and financial support) and the ONC as partner already on the ground (Beni/Nord-Kivu and soon in Boma and Isiro).

The ONC and the farmer trade union (Syndicat pour la Défense des Intérêts Paysans) have set up seven centres for the multiplication of rooted cuttings at Beni, Oïcha, Irango, Bingo, Mutwanga, Vuyinga and Mayi Moya in North Kivu. The Environment and Natural Resource Accounting Company has recently acquired an *in vitro* propagation unit to enhance the multiplication.

2.9.3. Use of variety resistance

The production of disease-resistant coffee plants is the most desirable method of achieving control of pathogens. It appears that the coffee varieties being grown currently are very susceptible to CWD. The revival of coffee research is necessary to increase productivity and offer coffee growers the option of using high-producing vegetative material. The effort has to be undertaken to obtain new CWD-resistant clones. The research task will have to be focused mainly on determining the behaviour and the level of resistance of the robusta clones (INERA elite clones and other materials introduced) towards CWD, as well as on trying to investigate the adaptation and introduction of highly productive material from foreign origins, such as clones S/2, 1S/3, 1S/6, 223/32, 257S/53 and 258S/24 from Uganda.

Researchers should also seek sources of resistance by prospecting in infected coffee plantations and obtaining germplasm that shows some level of natural resistance to the pathogen. Any promising material identified in the farmers' fields will need to be multiplied and tested to ensure that this material has a good degree of tolerance or resistance.

At the University of Kinshasa, coffee research is undertaken, and the University has taken part in various projects that have contributed to the improvement of an integrated management strategy for CWD. The study about varietal resistance was carried out in natural conditions of CWD at Beni in North Kivu.

Evaluation of resistance of these accessions under natural conditions has revealed differences in reaction in these accessions. One possibility now is to operate a preselection test for genetic resistance and to clone interesting material for the evaluation of resistance in the field. A future research activity is to study the level of resistance of the seedling material taken as survivors from devastated plantations.

To date, the results obtained (based on the mortality rate recorded on different genotypes 5 mo after inoculation) have shown that a certain level of variability exists ($P < 0.05$) in the coffee genotypes studied for their resistance to CWD

(Table 2.1). Different groups were identified by using Tukey's test. The materials tested were classified in four different groups: the first is composed of genotypes where the level of mortality ranged between 6.6% and 7.6%; the second, between 7.8% and 8.4%; the third, between 9% and 12.8%; and the fourth, between 13.3% and 16.1%. Considering these different groups, the first is represented by the genotype KR 8/10, where no infection was recorded, and the fourth is represented by the genotype KR 2/5, which was totally destroyed.

As part of the breeding for resistance studies, preliminary investigations were also made on the genetic diversity of the *F. xylarioides* population. An

Table 2.1. Varietal behaviour (percentage of mortality of inoculated plants) of different genotypes of robusta coffee tested for their CWD resistance.

No.	Genotype	% of Plant mortality	Group		
1	KR16/13A	6.6	a		
2	KR19/1B	7.5	a		
3	KR8/10	7.5	a		
4	KR19/11	7.6	a		
5	KR20/51	7.8	a	b	
6	KR19/18B	8.1	a	b	
7	KR17/55	8. 2	a	b	
8	KR10/7A	8.4	a	b	
9	KR18/10	8.49	a	b	
10	KR19/28	8.9	a	b	c
11	KR19/26	9.1	a	b	c
12	KR6/6	9.3	a	b	c
13	KR18/30	9.3	a	b	c
14	KR19/1A	9.5	a	b	c
15	KR19/55	9.7	a	b	c
16	KR20/50	10.2	a	b	c
17	KR19/31	10.3	a	b	c
18	KRA/6	10.6	a	b	c
19	KR16/55	11.0	a	b	c
20	KR17/47	11.2	a	b	c
21	KR1/3	11.2	a	b	c
22	KR16/13B	11.2	a	b	c
23	KR18/10A	11.4	a	b	c
24	KR1/1	11.5	a	b	c
25	KR19/18A	11.8	a	b	c
26	KR10/7B	12.1	a	b	c
27	KR19/12	12.3	a	b	c
28	KR20/31	12.4	a	b	c
29	KRC/3	12.5	a	b	c
30	KR3/5	12.6	a	b	c
31	KR12/6A	12.9	a	b	c
32	KR8/8	13.3		b	c
33	KR20/10	14.4		b	c
34	KR9/8	15.0		b	c
35	KR2/5	16.1		b	c

absence of genetic diversity within the population of *F. xylarioides* was revealed by molecular analysis using random amplified polymorphic DNA markers.

2.9.4. Sensitization of stakeholders

Extension is an essential part of management of the disease because farmers have to be informed of the results of research in order for them to be sensitized on the use of innovations. The structure of the Congolese coffee plantations must be taken into account when developing new strategies of extension and training of the farmers (e.g., the Farmer Field School approach). Financial, logistic and human means must also be taken into account, not only with the vast areas to be supervised but also, more especially, with the extent of the tasks to be achieved. Under the Common Fund for Commodities programme, ONC has conducted a training programme on CWD diagnosis and identification as well as training on management of the disease. We have initiated radio programmes in Beni and Isiro for farmers and all stakeholders, which cover all aspects of coffee production and CWD.

2.10. Conclusion

Coffee production in DRC should be flourishing because the ecoclimatic conditions in the country are favourable. However, coffee production is in decline because of the multiple factors that we have previously mentioned. Within the framework of revitalization, measures likely to improve the various levels of the sector such as production, processing, distribution and logistics must be considered.

The government of DRC, under the auspices of the multisector programme of rebuilding and rehabilitation, has initiated studies on the coffee sector. The ONC also has a plan of development for the coffee sector, which remains to be updated. It is important to say that even with limited support, the rural communities readily commit to the programme of development and take responsibility for themselves, often by organizing themselves into basic groups with the material and technical support of development non-government organizations or private companies; up to now, this has been largely independent of the government of DRC. Nevertheless, the government of DRC will have to continue to play an essential part in the development of the sector so that an effective development strategy for the revival of the coffee sector is produced. The government of DRC must encourage the private sector to intervene, restructure and regulate coffee-buying campaigns and support the local industrialization of finished coffee products.

According to all the studies made in DRC, it is important to continue to emphasize the threat and danger that CWD presents for the regions that remain currently free of CWD (provinces of Bandundu, Bas-Congo, Kasai Occidental and Kasai Oriental). Understanding the risk factors involved and alerting farmers to these risk factors, for example, rapid diagnosis of infected trees and that older plantations are more susceptible, as well as alerting them to the fact that improved propagation materials are available, will lessen the chance of CWD having such an impact in these provinces. In addition, it is important to put an emphasis on the risk of introduction of the disease into currently wilt-free areas by scientific staff in their survey inspection visits. Their equipment could be changed for every visit to reduce the chances of introduction.

Acknowledgements

We are grateful to the ONC coffee authority in DRC, the Kinshasa University, the Common Fund for Commodities and the EU for the generous financial support to the Laboratory of Phytopathology of the University of Kinshasa (DRC) during the implementation of this work. We would like to particularly thank Professor Kizungu Vumilia (Biometrics Unity of University of Kinshasa) for the contribution in the model logistic analysis. We also wish to express our gratitude to the Congolese stakeholders.

References

Banque Centrale Du Congo (BCC) (1995–2007) Rapports annuels d'activités. Service de Documentation.

Conjoncture Economique (2001) *Rapports annuals (1991–7)*. Ministère de l'Industrie, Commerce, Petites et Moyenne Entreprises et Artisanat, République Democratique du Congo.

Drachoussof, V., Focan, A. and Hecq, J. (1956) Le développement rural en Afrique Centrale 1908-1960/196fs2. *Synthèse et Réflexions* 1, 254–260.

Flood, J. (1996) A study of tracheomycosis or vascular wilt disease of coffee in Zaire. Report presented to Zairean Coffee Organization (OZACAF), August 1996, 13 pp.

Fraselle, J. (1950) Observations préliminaires sur une trachéomycose de *Coffea robusta*. *Bulletin Agricole du Congo Belge II* 2, 361–372.

Fraselle, J.V. and Geortay, G. (1952) Une grave maladie du caféier "Robusta" La trachéomycose. Avertissement et conseils aux planteurs. *Bulletin d'Information de l'INEAC I* 1–2, 87–102.

Fraselle, J.V., Vallaeys, G. and De Knop, O. (1953) La lutte contre la trachéomycose du caféier à Yangambi et le problème que pose actuellement cette maladie au Congo belge. *Bulletin Agricole du Congo Belge II* 6, 373–394.

Heim, R. and Saccas, A. (1950) La trachéomycose des *Coffea excelsa* et robusta des plantations de l'Oubangui-Chari. *Comptes Rendus des Séances de l'Académie des Sciences Paris CCXXXI* 11, 536–538.

Jagoret, P. and Descroix, F. (2002) Evolution de la culture de *Coffea canephora* en Afrique et problématique de développement. *Recherche et Caféiculture* 9, 44–59.

Kinds, A. (1930) Introduction d'espèces de Caféier Qu Congo Belge, *Bulletin Agricole du Congo Belge* XXI, 915.

Miny, M.P. (1930) La culture du Café Robusta au Congo Belge. *Bulletin Agricole du Congo Belge* XXI, 924–929.

Tshilenge-Djim, P., Kalonji-Mbuyi, A., Onyembe, P.M.L., Mukuna, K., Dibwe, M. and Oripale M. (1998) Caractéristique et évolution spatio-temporelle de la trachéomycose fusarienne du caféier robusta en République Démocratique du Congo (RDC). *Révue Congolaise des Sciences Nucléaires* 14, 132–140.

3 Coffee Wilt Disease in Uganda

G.J. Hakiza,[1] D.T. Kyetere,[1] P. Musoli,[1] P. Wetala,[1] J. Njuki,[2] P. Kucel,[1] P. Aluka,[1] A. Kangire[1] and J. Ogwang[1]
[1]*Coffee Research Centre, PO Box 185, Mukono, Uganda*
[2]*CABI Africa, UN Avenue, ICRAF Complex, PO Box 633-00621, Nairobi, Kenya*

3.1. Introduction

Uganda's economy depends largely on agriculture with coffee as the main foreign exchange earner. The contribution of agriculture to the gross domestic product has been as high as 51% in 1991 (Anon., 1993, 1994), but there has been a steady decline in the contribution to 36.3% in 2006 (Anon., 2006). This decline is not an indication of loss of significance of agriculture in Uganda's economy, but rather, it is due to the development of other sectors. Both arabica (*Coffea arabica* L.) and robusta coffee (*Coffea canephora* Pierre) are produced in Uganda. The distribution of coffee farming in Uganda is highly influenced by the agro-ecological adaptations of the two commercial species. Robusta coffee is grown at warmer and lower elevations, which vary in altitude from 800 to 1500 m in the central and southern parts of the country, covering total acreage of 240,000 ha. The annual rainfall in the robusta areas ranges from 1000 to 2000 mm. Arabica was first introduced in Uganda in 1900 from Malawi (then Nyasaland) for cultivation at all altitudes (Thomas, 1935; Leakey, 1970; Jameson, 1970), but these initial introductions failed to adapt to low altitudes <1400 m because of coffee leaf rust (CLR; *Hemileia vastatrix* Berk. and Br.) and climatic and soil conditions. Arabica cultivation is currently limited to areas within 1500 to 2300 m to the east, western and north-western Uganda, with an estimated total acreage of 37,000 ha. Rainfall in the arabica areas varies 1500 to 2000 mm annually. Both crops do well on deep fertile soils.

Most coffee farmers are smallholders (>90%) with an average hectarage per farmer of 0.23 ha for robusta and 0.36 ha for arabica. Both crops are grown predominantly mixed with food crops such as bananas and beans. The recommended spacing is 3 × 3 m for robusta and 2.74 × 2.74 m for arabica coffee. The national average coffee production varies between 400 and 500 kg/ha clean coffee (0.4–0.5 t/ha) with average weight ratio of cherries to clean beans of about 4.5:1 for robusta, and 300 to 400 kg/ha clean coffee (0.3–0.4 t/ha) with average weight ratio of 5.0:1 for arabica (UCDA, 2006). These yield levels are attributed to the fact that Uganda coffee is produced under low-cost systems. All arabica coffee is wet processed/washed, whereas robusta is dry processed.

3.2. Importance of Coffee in Uganda

The major export commodity for Uganda for many years has been coffee. Coffee contributes about 20% to 25% of the national annual foreign currency earnings from exports valued between US$84 million and US$456 million during the period from 1996 to 2005. More than 3 million Ugandans derive their livelihood directly from coffee as farmers, processors, exporters, transporters, traders etc. Arabica contributes 15% to 20% of coffee production by volume and 20% to 35% of the earnings from coffee exports (UCDA, 2005).

Although robusta is rated second to arabica coffee, it is an important commodity in the economy of many African countries including Uganda. It has a high extracting rate and therefore is best suited to processing into instant and decaffeinated coffee (De Graaf, 1986). In addition, robusta coffee generally has better resistance to CLR than arabica coffee. The bulk of Uganda's coffee is robusta (grown on 242,000 ha) and accounts for about 80% to 85% of coffee production by volume and raises 65% to 80% of revenue from coffee exports (UCDA, 2005). Robusta is most widely cultivated at altitudes ranging from about 800 m in Bundibugyo district (rift valley area) to about 1500 m, e.g. in Bushenyi district.

Robusta coffee (*C. canephora* Pierre) is indigenous to Uganda and is embodied in cultural and socio-economic norms of the Ugandan population. The Baganda used coffee beans in their ritual of blood brotherhood, and chewing of dried beans is still practised. Long before the arrival of the British explorers Speke and Grant in Uganda in 1862, the Baganda were growing robusta coffee around their homesteads (Thomas, 1947; Purseglove, 1968). The first coffee (robusta) exports from Uganda, mainly from Sese Islands (now Kalangala district) in Lake Victoria, were in 1878 (Thomas, 1947). From this humble beginning, robusta coffee cultivation expanded and thrived to become the major foreign exchange earner for the country for several decades. The importance of coffee in terms of export volume/quantity and value in US dollars, in comparison to other export commodities, is presented in Table 3.1. Although the volume of coffee exports has declined, it retains the highest foreign exchange compared to a wide range of commodities (Table 3.2).

3.3. Constraints to Coffee Production

There are a number of production constraints that confront coffee farmers. Among these constraints is a lack of varieties with resistance to coffee wilt

Table 3.1. Quantities of various commodities exported 2002–2006.

Commodity	Unit	2002	2003	2004	2005	2006
Coffee	t	201,591	146,299	159,983	142,513	126,887
Cotton	t	12,322	16,762	29,762	30,403	18,480
Tea	t	30,400	36,669	36,874	36,532	30,584
Tobacco	t	23,266	24,669	27,843	23,730	15,794
Maize	t	59,642	60,298	90,576	92,794	115,259
Beans and other legumes	t	10,753	18,070	26,233	28,332	27,087
Fish and fish products	t	25,525	26,422	31,808	39,201	36,461
Cattle hides	t	20,049	18,565	18,502	25,349	22,214
Electric current	000 kW h	264,685	217,486	193,104	62,577	53,019

Source: Uganda Export Promotion Board 2006.

disease (CWD), drought, other pests and diseases. Other constraints include low farm gate prices paid to farmers by coffee buyers or middlemen in addition to low and fluctuating world coffee price, which results in very small returns to the farmers. In the absence of stabilisation funds, many farmers have found coffee production increasingly less profitable. Moreover, there is no incentive to produce good quality coffee because there is no grading system to reward those who produce good coffee. Poor agronomic practises by farmers contribute to reduced productivity. For instance, stumping of coffee is one of the recommended practises for rejuvenation of coffee, but very few farmers practise it. The low adoption of recommended agronomic practises can be attributed to limited access to information/technologies due to inadequate extension services and to limited access to inputs. Inputs may also be available but are too expensive for most smallholders. The declining soil fertility is another concern because farmers do not use fertilizer or organic manure and is considered as the main cause of low productivity.

Table 3.2. Export commodities and their value ('000' US$) 2002–2006.

Commodity	2002	2003	2004	2005	2006
Coffee	96,626	100,233	124,237	172,942	189,830
Cotton	9519	17,755	42,758	28,821	20,474
Tea	31,293	38,314	37,258	34,274	50,873
Tobacco	45,262	43,042	40,702	31,486	26,964
Maize	10,609	13,724	17,896	21,261	24,114
Beans and other legumes	3284	5,235	8,968	8,693	8,162
Fish and fish products	87,945	88,113	103,309	142,691	145,837
Cattle hides	9810	4,925	5,409	7,064	8,030
Electric current	15,645	13,778	12,075	4,465	4,855

Source: Uganda Export Promotion Board 2006.

3.4. Most Important Diseases and Pests

Arabica coffee generally has more diseases and pests than robusta coffee, particularly at low altitude. Even at higher altitudes, arabica coffee suffers serious attacks of pests and diseases. Insect pests such as *Antestiopsis* spp., stem borers (*Bixadus sierricola* White) and leaf-sucking insects including leaf miners (*Leucoptera* spp.) and lace bugs (*Habrochila* spp.) can reduce yield and quality of arabica coffee. Incidences of coffee berry borer (*Hypothenemus hampei* Ferrari) have been observed to increase on arabica coffee at higher altitude than was previously seen. Root mealybugs (*Planococcus* spp.) are becoming increasingly destructive and difficult to control. The most damaging diseases are CLR (*H. vastatrix* Berk. and Br.) and coffee berry disease (CBD; *Colletotrichum kahawae* Waller and Bridge), which can cause yield losses of more than 50% under conditions favouring the disease. Bacterial blight (*Pseudomonas/Ralstonia syringae* van Hall) occurs in all arabica areas, but its occurrence is sporadic. Use of copper fungicides (50%) for the control of CBD and CLR at high altitude also keeps bacterial blight under control.

For many decades, robusta coffee cultivation in Uganda experienced only minor crop protection problems. The main pest on robusta has been coffee berry borer, which can be controlled by good cultural practises such as regular picking, stripping off and burning old and dry cherries ('mbuni') that remain on the trees at the end of the picking season. These practises remove sources of potential infestation and destroy the insects within the berries as well as their food source. Pruning is essential to keep the canopy more open, less humid and unattractive to the pest. Root mealybugs have been causing serious but localised damage on robusta coffee, and the symptoms have sometimes been confused with CWD. Other species of mealybugs and scale insects occur on aerial plant parts and can be a problem particularly on young plants just transplanted in the field or those still in the nurseries.

Diseases on robusta had often been minor and insignificant, for instance the berry red blister disease caused by *Cercospora coffeicola* (Cooke and Berk) has been present on robusta since the 1960s with occasional and sporadic incidences. Severe incidences of red blister disease causing berry fall have recently been reported in Rakai and several other districts. Research is needed on this disease to find effective management measures. Some robusta clones or varieties appear to have a lower resistance to CLR. Root rot caused by *Armillaria mellea* Vahl. occurs often on coffee established on land previously under forest but can also appear when shade trees such as *Ficus* spp. are suddenly cut down. Yellowing of affected plants followed by wilting and death can be mistaken for CWD.

The emergence of the highly destructive CWD in the early 1990s was a challenge to the whole coffee industry in Uganda (*Gibberella xylariodes* Heim and Sacc), of which *Fusarium xylarioides* Steyaert in the anamorphic phase is a highly aggressive disease that spreads rapidly, and all infected plants are killed. Whole or partial coffee fields were wiped out and, almost overnight, most farmers lost their only means of livelihood. Plates 1 and 2 illustrate the effect of the disease. This field was one of the most productive coffee fields at the Coffee Research Institute (CORI, formerly COREC), and within 3 to 4 years, it was reduced to a few unproductive trees.

More than 120 million robusta trees (44.5%) have been destroyed by the disease since 1993. In other words, 80,000 ha of robusta, 1.2 million bags (60 kg bags) of coffee and US$100 million lost to CWD (UCDA, 2004). In addition, many coffee farmers and other stakeholders who were dependent on robusta coffee have lost their livelihoods.

3.5. Brief History of CWD in Uganda

An account of the spread of the disease from the time it was first reported and the actions that followed are provided here and are meant to provide lessons learnt for the whole coffee sector and should allow for a better response to future epidemics.

In 1992, information was received from John Schluter, a businessman on a visit to the Democratic Republic of Congo (DRC), of a devastating robusta coffee disease in the Beni and Isiro areas of DRC and warned of its consequences if allowed to cross into Uganda. To find out as much information as possible on the new disease, its symptoms, causal organism, epidemiology and control, the senior author turned to literature to be able to recognize the disease and what to do about it should it appear. Recommendations for control of the disease were obtained from literature (Wrigley, 1988; Coste, 1992). Uprooting and burning of infected plants at the earliest symptoms and other sanitary measures were intended to destroy sources of infection/inoculum to reduce the rate of spread within the field and outside the field. Infected plant parts carried by people as firewood was a mechanism for disease spread as were agencies such as wind or rain. Spread of the disease through these agencies ensured dissemination far from the original affected farms.

Ironically, in September 1993, wilting and death of a few robusta coffee trees were observed in a 2.8-ha experimental plot at CORI, Kituza, in Mukono district of central Uganda. The plant pathologist (main author) recovered *F. xylarioides*, the anamorphic form of *Gibberella xylarioides*, from infected plant parts (stems and roots). Identification of the pathogen was based on cultural and spore morphological characters as described by Booth (1971) and Gerlach and Nirenberg (1982). Reproduction of symptoms on seedlings was done by transplanting 6-month-old seedlings/clones in sterile soil infected with spores of the pathogen and maintaining the inoculated plants in the screen house until symptoms appeared. Re-isolation of the pathogen from infected seedlings confirmed the pathogenicity of the fungus causing the wilt.

The recommended standard control method for CWD is uprooting and burning of the diseased plants to eliminate potential inoculum sources for further infection and was maintained in this particular plot from 1993 to 2000. At first, the method appeared effective. However, with time, the method proved unsustainable due to the high cost of labour for uprooting the diseased plants. It appeared that inoculum sources were not only from within the same field but also from external sources. It transpired that coffee trees in the adjacent forest were also infected by wilt, and it was suspected that this was the source of inoculum. As long as other sources of inoculum exist, from outside the targeted field, infection continued. Farmers should therefore work in conjunction with their neighbours to minimize the transfer of inoculum from one farm to another.

Plants can be affected at any stage of development, from seedlings to adults. The first symptom exhibited from the top of infected plants is a flaccidity of the leaves on one side of the affected stem. Chlorosis may or may not occur. Wilting, rapid defoliation and dieback soon follow (Plate 3). During the rainy season, leaf wilting may not be readily observed, and the only indication of wilt is rapid defoliation (noticed as carpet of leaves at the base of trees) and dieback, which is the progressive death of branches or shoots beginning at the tips leaving behind bare dry twigs. Dieback at the tips of the main stem spreads down to all primary branches resulting in death of stem. Dark brown necrosis occurs on stems of young suckers and branches, which leads to death of those parts. Leaf veins show dark brown necrosis. On a multistemmed coffee plant, stems die in sequence, one by one until the whole bush dries up completely. Coffee berries on the affected tree ripen prematurely (turn red) and dry up but remain attached to the primary branches.

CWD is confirmed by the presence of blue–black streaks or bands in the wood when a bark of the affected stem is peeled off. This distinguishes CWD from all other wilting, which could be due to root mealybugs *A. mellea* and other fusaria such as *Fusarium oxysporum*, *Fusarium solani*, *Fusarium lateritium* and *Fusarium decemcelulare.* At the advanced stages, cracks may occur on the stem from around the collar region and up along the stem. Within these cracks in the bark are embedded brown to black perithecia, which bear numerous ascospores. Ascospores are the sexual or teleomorph phase of the fungus. These are spread by wind or washed by rain into soil to become soil inoculum, then spread by runoff to other plants in the vicinity.

Infected plants do not recover even after pruning and/or stumping. When pruned or stumped, any suckers that may sprout develop dark brown necrosis and eventually wilt, die and dry up.

The affected and dried up plants remain firmly rooted in the ground, unlike those affected by fungi such as *Armillaria* or root mealybugs, which easily topple over when pushed.

Following the outbreak in Mukono, recommendations were made based on literature (Coste, 1992) as an emergency measure in the absence of any other information and were not then based on research findings done in Uganda at the time.

Thus, regular field inspection to detect early symptoms of the disease was recommended to farmers. Any plants confirmed as infected were to be uprooted and burnt on the spot. Farmers were advised to avoid using coffee husks as mulch in their coffee fields because these were strongly suspected to be contaminated with the pathogen. Some districts, e.g. Masaka, passed a by-law prohibiting milling of coffee from outside the district. Restriction on the movement of infected plant materials, coffee husks and soil was also recommended. The berries on infected plants were to be separately picked from coffee on healthy plants. Farmers were also advised to plant coffee on new land, preferably some distance from the infected field. Replanting in the affected fields was recommended not be done immediately after uprooting but to delay planting for at least 1 year and preferably longer; other crops can be grown. Tools were to be cleaned by flaming or use of disinfectant to avoid passing infection from an infected plant to another. When pruning or handling plants, always start with healthy-looking plants first and sterilize tools after each plant. Wounding the stems during cultivation (such as weeding) of the coffee should be avoided because wounds are likely sites for entry of the pathogen into the plants. These recommendations are now backed by research findings that have been conducted in the past 4 to 5 years during the project period, and therefore, the above management measures remain.

The first report of wilt outside Mukono district was received in October 1993 from Bundibugyo district in south-western Uganda bordering the DRC concerning the presence of a destructive wilt disease of robusta coffee. The Uganda Coffee Development Authority (UCDA) sponsored a multidisciplinary team of researchers composed of a pathologist, an entomologist, an agronomist and an economist to travel to Bundibugyo and identify the problem. The main author was part of the expedition. The affected robusta areas were Bwamba, Bubukwanga and Bubandi counties. In this district, both robusta and arabica are cultivated. Arabica coffee occupies higher grounds in Nyankonja, Busaru and parts of Bubukwanga counties. The disease was reported only on robusta coffee. During the survey, two farms in each of the three subcounties were visited, and samples were collected. All the fields visited had some dead plants and partially diseased trees. Symptoms of the disease were not typical of the wilt disease caused by *F. xylarioides* as described in literature (Coste, 1992) and observed at Kituza hardly a month earlier. The dead and partially diseased plants had dieback, some leaf fall occurred but a lot of dry leaves remained attached to the dead branches. When bark was scraped off from the stem of the affected plants, there was no blue–black streak characteristic of *F. xylarioides*, instead only brown to dark brown disintegrating/rotting tissues were observed. A whitish powder covered most of the dead and blackened suckers and branches in many trees. From the samples collected, the fungi recovered in the laboratory were *Fusarium stilboides*, *F. lateritium*, *F. solani* and *F. oxysporum*, which can also cause wilting and death of plants under stress. In addition, there was a high incidence of stem borer attacks leading to wilting and death of trees in the same fields. During this first visit, the presence of CWD was somehow missed probably due to the high incidence of other fungi and stem borers. It was later that these other fungi became less frequent, and *F. xylarioides* became most dominant from specimens received from this area (Hakiza, 1998, unpublished report).

More reports and samples were received in 1994 from Kanungu district (formerly part of Rukungiri) in western Uganda, and the CWD pathogen was consistently recovered from samples of roots, stems and branches/primaries. Specimens from the Bundibugyo district, received about the same time, also clearly revealed the presence of the pathogen.

The worsening situation prompted UCDA in 1996 to organize and fund a joint survey conducted by a multidisciplinary team from the Ministry of Agriculture Animal Industry and Fisheries (MAAIF), Coffee Research Centre (COREC) and UCDA to establish the distribution/extent and damage caused by the disease in ten major coffee-growing districts of Uganda. It revealed the presence of the disease in ten districts: Mukono, Mubende, Mpigi, Ntungamo, Kasese, Kabarole, Rukungiri, Bundibugyo, Kiboga and Masaka.

In the same year, UCDA invited Dr Julie Flood of CABI to assess the situation, as she had just conducted a survey in Eastern DRC (Isiro area) on the same problem (Flood, 1996), and she made an assessment in January 1997. Her findings confirmed the presence of CWD as had been reported by the national scientists. She recommended regional collaboration to enable Uganda to benefit from the past experiences of African countries that had previously successfully managed outbreaks of CWD in earlier decades.

A regional workshop on CWD was held in the International Conference Centre, Kampala, from July 28 to 30, 1997. Representatives from the International Coffee Organization (ICO), CABI and CIRAD attended the meeting, as well as participants from African countries. The main outcome of the meeting was the initiation of a regional research project proposal (Regional Coffee Wilt Programme) to contain the situation. Uganda and DRC, which were the worst affected countries, were to play leading roles in research and development on CWD. This was a major landmark in the struggle against CWD in the region. The project was finally initiated in 2000.

However, the incidence of the disease had continued to increase through 1995–2000, and samples and reports were received from other parts of Mukono district, Kiboga, Mpigi, Jinja, Iganga and Mayuge. In 2000, reports and samples were received for robusta coffee from the West Nile region where both coffee species are grown; the pathogen was reisolated. By the end of 2000, all robusta coffee districts in Uganda had been affected. Consequently, during the intervening time between the research proposal (Regional Coffee Wilt Programme) being written and obtaining the funding, the National Agricultural Research Organisation (NARO) of Uganda became so concerned that they convened a technical workshop on CWD in 1999 to examine research progress and define action points to accelerate research activities toward finding a solution. Following this meeting, a task force composed of plant pathologists, plant breeders and socio-economists from NARO and Makerere University was formed to formulate a research programme on CWD and to decide on its implementation. A comprehensive research programme was drawn up for a period of 5 years together with its budget. CORI is continuing to implement the programme.

Many reports of wilt on arabica coffee and samples received at different times from all the arabica areas in the eastern, western and north-western/West Nile region of Uganda revealed other Fusaria but not *G. xylarioides*. Many follow-up visits to those areas have also been made to investigate and verify the reports. In all cases, *G. xylarioides* has not been recovered from samples or observed in any of the field visits made. The fungi that have been consistently recovered from samples were *F. solani* and *F. oxysporum* and were confirmed as the cause of localised wilting and death of arabica coffee in those districts (Arua, Nebbi, Bushenyi, Kasese, Ibanda, Rukungiri, Mbale, Sironko, Bududa and Manafa districts). Root rot caused by *A. mellea* Vahl has also caused concern and has been mistaken for CWD. Insect attacks, the most common of these encountered in the arabica areas, that induce wilt symptoms that can be confused with CWD are coffee stem borers (*B. sierricola*), coffee root mealybugs (*Planococcus ireneus*) and twig borers. In essence, to date, CWD infects only robusta coffee, even in districts where the two crops are grown side by side.

The absence of the disease in arabica coffee in Uganda shows that the pathogen is currently host specific. It is not known if with time the pathogen will change and infect arabica as well, particularly in areas where the two species are grown side by side. *G. xylarioides* remains a potential threat to arabica coffee in Uganda.

3.6. Coffee Wilt Distribution in Uganda

In Uganda, CWD infects robusta coffee (commercial species), but it also affects *Coffea liberica* and *Coffea kapagota* in the germplasm collections at Kawanda Agricultural Research Institute. The disease manifests itself as classical vascular wilt with wilting leaves sometimes preceded by chlorosis followed by dieback. A characteristic blue–black staining of the wood is seen under the bark of affected trees.

After 1993, the incidence and severity of CWD continued to rise causing decline in yields in all affected districts. The worst affected districts were Bundibugyo, Kyenjojo, Hoima, Kabale and Kayunga, where it was estimated that 40% to 50% of the coffee fields were completely destroyed or abandoned (CORI, 1996–1997; Lukwago and Birikunzira, 1997; UCDA, 1999). An estimate by UCDA in 1999 indicated that 14.5 million robusta coffee trees (4.8% of the coffee) were destroyed countrywide. However, these estimates were based on inadequate information from an earlier survey conducted in 1996 in only ten districts. The data capture tools used in the 1996 survey were not comprehensive. Moreover, there were no follow-up farm visits to monitor and check the seemingly rapidly increasing rate of disease spread and incidence levels. Consequently, a more accurate estimate of current magnitude of the CWD problem was needed to provide quantitative information on the current status of CWD in the country. Various surveys (biological, remote sensing and socio-economic surveys) were therefore conducted to provide comprehensive and up-to-date information on the status of CWD in Uganda in 2002 and again in 2004.

The first biological survey was carried out in Uganda in March to April 2002 with the following objectives:

- To establish CWD prevalence and severity at farm, district and national levels
- To quantify the impact of CWD on coffee production at farm, district and national levels

- To assess the influence of human and biophysical factors within the coffee farming system on incidence, spread and severity of CWD in Uganda
- To develop a survey tool for biological surveys in African coffee-growing countries participating in the Regional Coffee Wilt Programme (Improvement of Coffee Production in Africa through the Management of CWD).

The survey was conducted in ten districts in the central, six districts in the western and five districts in the eastern part of the country. The districts surveyed in the central zone included Mukono, Kayunga, Wakiso, Mpigi, Luwero, Masaka, Rakai, Sembabule, Mubende and Kiboga. The coverage in central Uganda included almost all coffee-producing districts except the Lake Victoria district of Kalangala and the marginal rainfall district of Nakasongola, which are minor producers. The ten districts selected from the Central region produce 80% to 90% of robusta coffee in the country. In the West, the districts surveyed included Hoima, Bundibugyo, Kibale, Kyenjojo, Bushenyi and Rukungiri. Similarly, in the east, throughout the period from 1995 to 2000, reports of the disease were received from other parts of Mukono district, Kiboga, Mpigi, Jinja, Iganga, Mayuge and Iganga; the West Nile region was affected in 2000. Thus, by the end of 2000, all robusta-growing coffee districts in Uganda had been affected (Fig. 3.1).

In each selected district in the central zone, a number of robusta-coffee-growing parishes were randomly selected. The number of parishes selected from each of the districts was proportional to the total no of parishes in the districts. In the western and eastern zones, two parishes were selected from each selected district. Three villages were randomly selected from each of the selected parishes except in Mukono and Kayunga where one village was randomly selected. In each village, three farms were randomly selected except in Mukono and Kayunga where nine farms were selected. Data were collected from each of the selected farms using a data input form.

Survey results indicated that CWD was present in all 21 districts, although at varying incidences. All farms surveyed in Bushenyi, Hoima, Jinja, Kamuli, Kibale, Kiboga, Kayunga, Luwero, Mukono and Rukungiri districts had CWD. Rakai district had the highest percentage of farms without CWD. These results were based on the assumption that the missing coffee trees on the farms were assumed to have been infected by CWD and destroyed by the farmers as recommended.

Only Iganga had a low mean CWD incidence of 3.5%. Bugiri, Rakai, Mayuge, Sembabule, Masaka and Bundibugyo had a moderate percentage of incidence of CWD. Bundibugyo is where CWD was first reported, and farmers there had been advised to uproot all affected trees as soon as symptoms appeared. CWD incidence in the remaining districts was very high. Kayunga, Luwero, Kibale, Mubende, Hoima, Mukono and Kiboga were most affected with more than 60% of the trees infected (sick or dead-stumped or dead-standing or uprooted due to CWD). The severity of CWD in farms was high, ranging from 3.5 in Iganga to 5.6 in Bundibugyo on a 1–6 scale and ranged from 1.1 in Iganga to 3.3 in Kibale.

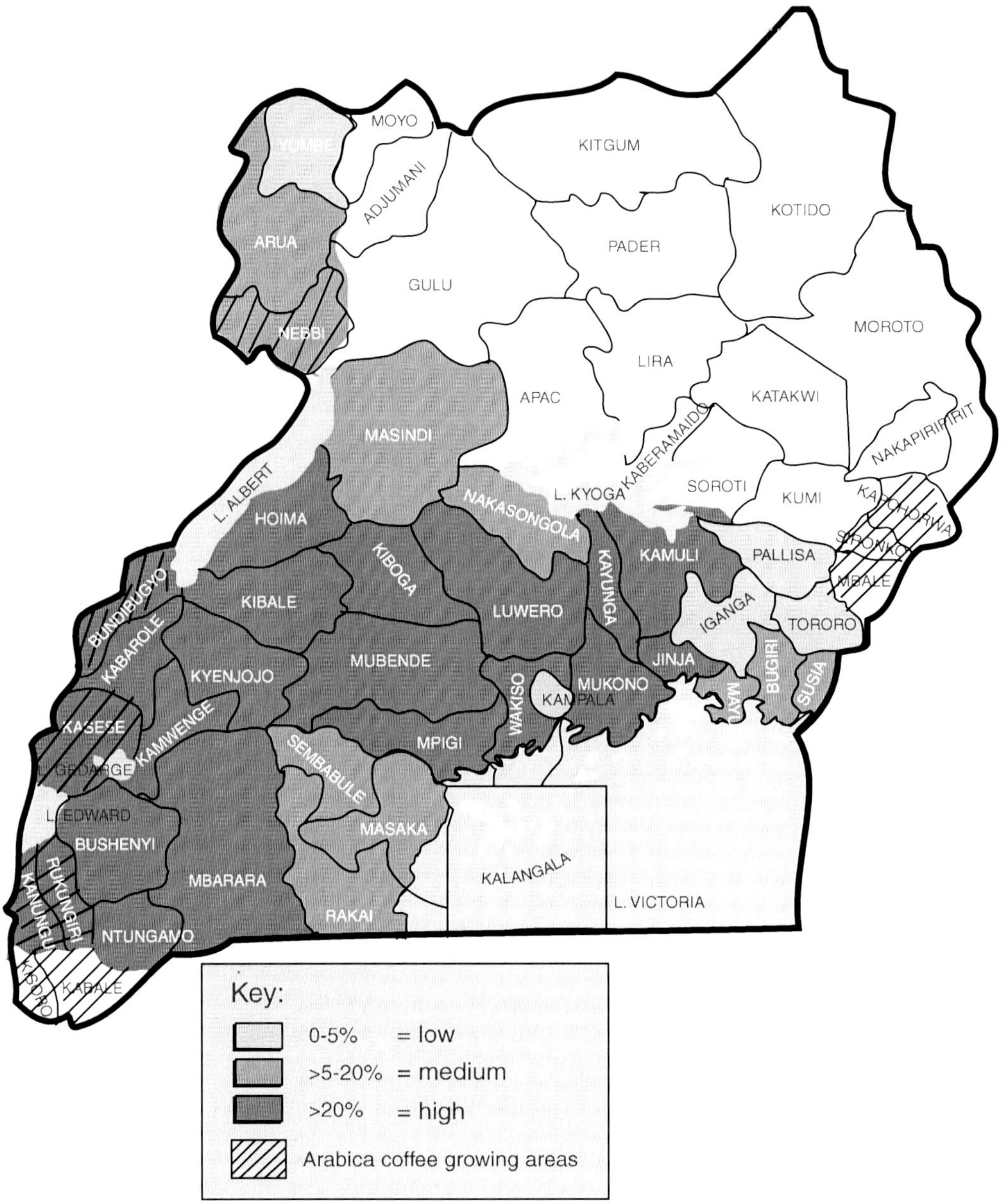

Fig. 3.1. Map of Uganda showing the distribution of CWD of robusta coffee (not to scale).

3.7. Importance of CWD in the Country

Robusta coffee is a key commodity in household incomes and Uganda's main revenue earner. The disease has serious implications to the national economy as a whole but even of more concern to the household income of the farmers. For example, the farmers visited in 1999 in the Mukono, Rakai and Mpigi

Table 3.3. Loss of income incurred by two farmers due to CWD in Mukono district.

	Case 1		Case 2	
Period	Period	Case 1	Case 2	Income (gross revenue in Uganda shillings)
Before CWD	5600	3,920,000	2100	1,470,000
1 year after CWD	2800	1,960,000	420	294,000
2 years after CWD	0	0	0	0

Source: Field visits to selected farmers in Mukono district (CWD Task Force Report, 1999).

Table 3.4. Yield loss in 60-kg bags of dry cherry (kiboko) due to CWD (inclusive of missing trees).

District	Average annual yield (kg) before CWD (B)	No. of farms giving value for yield before	Average current annual yield (kg) (after CWD) (C)	Farms giving value of current yield (after CWD) (C)	Average % yield loss (current–before)	Farms where % loss could be calculated
Bugiri	18.0	17	7.4	17	58.9	17
Bundibugyo	16.1	18	2.3	15	85.7	15
Bushenyi	22.5	17	2.6	16	88.4	15
Hoima	16.1	18	0.9	18	94.4	18
Iganga	12.9	17	3.4	15	73.6	15
Jinja	18.6	18	4.3	17	76.9	17
Kamuli	20.0	17	6.1	17	69.5	17
Kayunga	46.5	63	3.4	57	92.7	57
Kibaale	25.0	18	2.4	18	90.4	18
Kiboga	41.6	50	4	50	90.4	49
Kyenjojo	16.3	18	2.1	18	87.1	18
Luwero	32.4	134	4.1	133	87.3	130
Masaka	17.4	122	7	118	59.8	114
Mayuge	7.5	15	3.7	15	50.7	15
Mpigi	26.8	134	4.5	128	83.2	123
Mubende	27.2	140	2.9	133	89.3	129
Mukono	40.0	140	3	141	92.5	128
Rakai	14.9	115	9.8	109	34.2	106
Rukungiri	15.6	18	3.3	17	78.8	17
Sembabule	33.5	31	19.1	30	43.0	30
Wakiso	17.7	139	2.2	128	87.6	127
Grand total		1259		1210		1175
Mean	26.2		4.7		76.9	

Source: Report on CWD Biological Survey, 2002.

districts reported that the disease had adversely affected their earnings and subsequently their ability to meet household cash requirements such as school fees, domestic needs, hired labour, taxes and investment (CWD Task Force Report, 1999). According to the same report, the estimated average loss in income due to CWD per household was 362,973 Uganda shillings per annum, which was well above the average per capita income of Uganda estimated at US$250. Members of the task force visited some farmers in Mukono district who had incurred losses due to CWD. Two of the farmers who kept records revealed their losses as indicated in the table below.

Within a very short time, the disease caused a total loss of the crop. Many farmers, particularly those who invested substantial amounts of money to improve their ten or more hectares of coffee, did not only lose their crop and income (Table 3.3) but also their health.

Yield loss at a district level (Table 3.4) showed that CWD was causing 77% loss in yield of robusta coffee at the national level. The mean yield loss across districts ranged from 34.2% in Rakai district to 94% in Hoima. Even among districts with the least losses in yield such as Rakai, some individual farms incurred losses of more than 50%.

The current situation indicates the presence of the disease in all robusta districts (Fig. 3.1). Its presence in some non-traditional robusta-growing areas, e.g., Busia and Arua districts, has also been confirmed. The incidence and severity, however, have markedly declined in all the districts probably because the most susceptible plants were attacked first and those that have escaped or have some resistance are currently being infected. Farms that had low incidences during the 2002 survey, e.g. Rakai and Iganga, are currently experiencing some increase in incidence of CWD. Despite the low incidence of the disease in some districts, farmers and all stakeholders should maintain vigilance and conduct disease management practises as has been recommended. Under favourable conditions, CWD could re-emerge.

3.8. Factors Affecting Severity or Incidence of CWD in Uganda

There are probably many factors, some known and many still largely unknown, that influence the incidence and severity of CWD. Some of the factors are examined here.

3.8.1. Lack of information

One of the factors that could have accelerated the spread of the disease in Uganda was lack of information/knowledge on all aspects of the disease. Research, extension and the Ministry of Agriculture had no useful information to give to farmers to slow down or stop the epidemic. In addition, it was widely believed that robusta coffee was resistant to most pests and diseases. This state of uncertainty gave time for the disease to continue to spread to all districts within a relatively short time.

3.8.2. Varietal susceptibility to CWD

Most robusta plantings in Uganda originated from seeds selected by farmers from their own coffee fields. Farmers preferred their own seeds/seedlings, although there are recommended clones or clonal seeds available that have better yields, resistance to some diseases and good bean and cup qualities. The dependence on self-saved seeds/seedlings was largely because they incur no cost in obtaining them whereas the recommended clones/varieties were very expensive. The cost of recommended robusta cuttings was 500 Uganda shillings for each single clone, and seedlings cost 300 Uganda shillings each. Clonal seeds from the research station cost 2000 per kilogram. In addition to cost, the source of good planting materials may be far removed from the farmers, so further costs in transportation are incurred; all these costs encourage farmers to use their own materials. This seedling coffee was very susceptible to CWD and encouraged the spread of the disease. Although CWD can also infect the recommended clones, partial resistance is evident as compared to seedlings.

3.8.3. Environmental factors

At the time of the onset of the epidemic, rainfall was observed to be higher than normal for a prolonged period of time. Favourable environmental conditions, in the presence of susceptible germplasm and an aggressive pathogen, led to the disease spreading rapidly throughout the country. Two rainy seasons normally occur in Uganda. The first rainy season (March–June) is normally heavier than the second rains (September–November). Recent field observation trials at Kituza revealed an increase of CWD and disease progress in infected plants during the rainy season and a drop in number of new infections during the dry season. This can be attributed to increased plant activities during the wet season, which enhances fungal infection and colonization of the host as well as rain being involved in the spread of the disease.

In addition to climatic factors, soil type seems to have an influence in that farms on predominantly sandy soils had more wilt-infected coffee trees than farms on predominantly loam, sandy–clay or clay soil types.

3.8.4. Lack of preparedness and resources

Generally, the response to the epidemic was slow for various reasons. Because it was new, no one could predict the magnitude or gravity of the problem. Lack of resources and lack of preparedness to mitigate emergency situations such as this could not allow a quick response. Despite recommendations for uprooting and burning, as well as other sanitary measures, farmers lacked resources to uproot and burn infected plants. The exercise proved too expensive and too demanding on the ageing farmers who are the majority. Lack of resources also delayed the start of serious research by the national research institutions. Research priorities were based on constraints with measurable impact on the farmers and national economy but could not readily accommodate emerging problems. Lessons learned from these experiences probably caused a more positive action when bacterial wilt of bananas (*Xanthomonas musacearum*) emerged in Mukono district in 2002 (Banana Bacterial Wilt Task Force Report, 2002, 2003). Mobilization of farmers for uprooting and burying infected banana stools was quickly done and a research agenda defined and funded within the same year the report was received, which was in total contrast to what happened when CWD was first observed.

3.8.5. Human activities

Unknowingly, farmers, farm workers and others spread the disease within the farms and out of the farms to their neighbours and to other districts. Within the farms, farmers and others cut the dry trees for use as firewood at the homestead. The mode of transport of the dry plants was by dragging them through the coffee field and taking the shortest route to the homestead, to where these were stacked outside the kitchen house. Shortly afterwards, more infected trees were observed along the routes taken through the coffee field. The infected stems were also transported on bicycles and through other means of transport to distant places where new infections could have been started.

G. xylarioides is considered a wound pathogen, which can only invade the plant through wounds or damaged tissue (Muller, 1997). In many normal farming practises, such as weeding using hoes or slashers or machetes/pangas, wounds can be created on roots and stems. A higher disease incidence was evident when coffee was hoe weeded as compared with other weed management practises. Pruning and desuckering also create wounds, which provide entry points for the pathogen, and the incidence was greater in coffee where the canopy was unpruned as compared to coffee under canopy management (pruned). High disease incidence was also observed on farms that did not use fertilizer or manure. Wounds created close to stem base are more likely to result in systemic infection than wounds further up the plant (Hakiza *et al.*, 2004). Under favourable conditions of temperature and moisture, spores of the pathogen deposited on the wounds germinate and penetrate the host resulting in infection, and later visual symptoms of the disease would appear. Shading of the coffee increased the probability of disease incidence, and coffee farms with shade trees had significantly more infected trees than farms without shade.

Increased CWD incidence was also observed on farms where livestock were tethered to coffee trees or animals were allowed to graze in coffee fields, as compared to farms where livestock (cattle, goats, sheep and pigs) were not tethered or allowed to graze in coffee fields. Extensive wounding of stems was observed as well on big roots where pigs had been digging up soil. Most probably these wounds led to increased infection.

Infected planting material/seedlings could have played a role in the dissemination of the wilt pathogen. It is widely believed that CWD crossed to

Jinja, Mayuge and other districts in Eastern Uganda, mostly through volunteer seedlings collected from forests in Mukono district where CWD was also rampant on forest coffee. The farmers are attracted by the low cost of these seedlings compared to good seedlings from known sources. The seedlings may appear healthy but could be infected or contaminated with *G. xylarioides*. Symptoms may develop later when transported to other areas for planting. Abundant inoculum in the form of perithecia is readily produced around the collar region and a few centimetres below the soil surface following the death of the seedlings. This inoculum is washed by rain into soil or spread by wind to initiate infection to healthy seedlings planted in the same field or further.

As a soil-borne pathogen, any material that carries soil is suspect. Nursery operators often collect topsoil from the forests, as this soil is considered very fertile. It has been demonstrated that unsterilized soil collected from forests may have enough contamination with the disease pathogen to cause infection to seedlings (Flood, 2005). The seedlings/clones raised in soil contaminated by the wilt pathogen can be transported far and wide and cause new infections when they die, and perithecia are formed abundantly around the collar and just below the soil surface.

Coffee plantations had significantly more CWD than other production systems, and coffee husks were used for mulching in many coffee plantations. A number of farmers who lost their coffee have attributed this to mulching with coffee husks. It was also claimed that the disease got to Mukono in coffee brought from the infected districts in western Uganda and hulled in Mukono district. Many farmers took the husks for mulching their coffee fields. Mukono district is among the districts that had a high incidence of CWD. Because the coffee husks were suspected carriers for the pathogen, its use for mulching was stopped by the local leaders. The recommendation is maintained to date.

3.9. Country-Specific Management Practices in Uganda

3.9.1. Farmer support and other establishments

Control of CWD requires the participation of all coffee stakeholders from village level to policy makers. Each stakeholder has a role to play to ensure the effectiveness and success of control programmes. On-farm measures to prevent spread of the disease are done by the farmer himself, ideally in conjunction with his immediate neighbours, and good sanitation measures can minimize the transfer of inoculum from one field to another.

Key coffee stakeholders spearheading the fight against CWD comprise MAAIF, which is responsible for providing extension services to farmers, policy issues and provision of funds. Non-government organizations, such as World Vision, also do some extension work, as well as various farmer organizations, among them the National Union of Coffee Agribusinesses and Farm Enterprises. The UCDA is a body created by the government to oversee the coffee subsector and to provide extension services to farmers through its extension agents posted to major coffee production districts. Research services are provided mainly through NARO, universities and other private organizations. NARO, through its CORI, conducts research to solve priority constraints of which CWD has been of major economic concern.

Coffee research in Uganda has been directed towards the control of CWD to increase the knowledge base of the disease and its pathogen. The formulation of effective means of control/disease management and breeding for durable resistance to CWD depends on detailed knowledge of the pathogen. Concurrently, research is also directed to finding agronomic measures for management of the disease. Use of resistant cultivars is the most effective long-term measure against CWD. In addition, resistance can eliminate or reduce the need for other measures. General selection and breeding for resistance have been making use of available germplasm in the coffee collection at Kawanda Agricultural Research Institute and from survivors of CWD in the farmers' fields. The goals of these include improvement of yield and vigour, better local adaptation (resistance to other pests and diseases) and better bean quality. There are high expectations from the selected CWD-resistant robusta coffee lines, which are undergoing final field evaluations on farms in several districts before release to farmers on a large scale (see Chapter 9).

3.9.2. Replanting programme in areas infected by CWD

Coffee production in Uganda had been on the decline for over a decade as evidenced by the reduced volume and value of the most important export commodity (Table 3.1). Among the reasons for the decline was low productivity due to old trees (more than 50 years old). To revitalize production, a gradual replacement of old robusta trees with new plantations, using the six high-yielding clonal robusta, was planned before the emergence of CWD. The replanting programme was initiated in 1990 by UCDA, and when the disease emerged, this operation continued and has been effective in maintaining coffee stands and ensured that there was some production of robusta coffee.

Replanting has been based on the recommended clones in the absence of better varieties that could be used. As a precaution against CWD, all the nursery operators were trained in nursery management, CWD control and general coffee field management including control of other pests and diseases. The planting materials were purchased by UCDA and distributed free of charge to farmers to encourage them to replant and also to ensure they get the recommended planting materials. It has been estimated that 100 million clonal seedlings were supplied to farmers since the start of the programme in 1990 (UCDA).

3.9.3. Introduction of robusta coffee in non-traditional areas free from CWD

To avoid/escape CWD, the production of coffee in selected areas suitable for robusta coffee, outside the areas where robusta had been traditionally grown, was encouraged and supported. Some areas in the north and east of the country were already growing some coffee. With UCDA support and technical input from the Coffee Research Centre, training of extension workers and nursery operators was undertaken. Training included general coffee management with an emphasis on CWD especially to avoid spread of the disease from infected districts to new areas. The programme of capacity building and raising awareness is ongoing.

3.9.4. Promotion of arabica production at lower altitudes (1400–1500 m)

Because CWD has not been observed on arabica coffee and also confirmed through inoculations that it is not infected by the pathogenic strain present in Uganda, it seemed logical to promote the cultivation of arabica in the robusta areas where CWD is prevalent. Trials were initiated using three Catimor varieties, which are resistant to CLR but highly susceptible to CBD and cannot be grown successfully at high altitude. The quality profiles of the varieties at various altitudes have been assessed, and results show that some areas 1400 to 1500 m could produce good-quality arabica.

3.9.5. Training of trainers

The need for information on the managements of CWD and the dissemination of this information has been high. In order for farmers and extensionists to manage the disease, they needed information from research. Lack of information at the beginning allowed the disease to spread rapidly. As information became available, there was need to increase information flow between farmers, researchers and the extension systems to manage the disease more effectively.

Under the CFC-funded project, CORI at Kituza together with their partners CABI conducted a Training of Trainers in some disease recognition and management based on available information. CORI, in conjunction with CABI Africa, trained 75 extension staff of the MAAIF and an additional 25 extensionists belonging to the National Union of Coffee Agribusinesses and Farm Enterprises coming from all coffee regions including arabica areas and from those areas still free from the disease These later groups were trained in the recognition of disease symptoms to allow early detection of the disease and hence improved surveillance. This is crucial for a quick response to enable effective control to be applied before the disease is widespread. These extensionists now form a basis for farmer training on CWD management.

3.9.6. Participation of leaders in the CWD control programme

Under the decentralized system of governance, the districts operate as small units with power to formulate and implement policies affecting their areas and to recommend to the central government issues which require new legislations. It was observed that leaders in the districts would be useful partners in the campaign against CWD and therefore should be sensitized on the problem. The aim of sensitization was for them to appreciate the magnitude of the problem and its effect on the coffee farmers and the local communities in general. It was also essential to enlist their participation in sensitization and mobilization of farmers to try to manage the disease particularly where by-laws are required. Successful sensitization workshops for local leaders have been carried out in ten districts to date and are ongoing and have allowed them to influence their respective communities and mobilize and sensitize farmers effectively.

3.9.7. Information dissemination

Through mass media

Radio programmes in Luganda for farmers and other stakeholders go on air every Sunday morning at 8:30 AM by UCDA on Radio Simba/CBS. Information on all aspects of coffee production, crop protection in particular CWD and coffee trade are covered in the programme. The programmes have been sustained for more than 2 years now, covering about 30,000 listeners each time the programme goes on air.

Farmer field schools approach to information dissemination

The traditional extension service is where the service provider/extension worker is the messenger between research and farmer (Ladela, 2001). The farmer is the passive recipient of this information of which he has never taken part in its generation; consequently, many good recommendations have never been adopted. However, farmers are known to possess a lot of information, which they have acquired over time through experience (Marseden, 1994). Farmers too have the ability to analyse situations and make rational decisions.

Information dissemination through the traditional extension system is very slow because the number of extension workers is quite low compared to the number of farmers, whom they deal with one by one. To enhance the farmers' ability and to accelerate information dissemination, a farmer field school (FFS) approach was initiated. Through participatory approaches, farmers work in groups and have the advantage of learning from each other as they share information. In an FFS, farmers make regular field observations, relate their observations to the ecosystem and apply their previous experience and any new information to make management decisions on their crop guided by the extension worker.

A coffee FFS follows the production cycle of the crop. It consists of a group, usually 20 to 30 farmers, who set up a study field. The group is responsible for the care and maintenance of the study field from soil preparation to harvesting/post harvest. Such schools have been started in

Bugiri, Kayunga, Mukono, Kiboga, Masaka and Rakai districts. Each of these districts has at least two active field schools. Information dissemination through FFS has been so successful through open field days hosted by these farmers, so that in all the six districts, extension workers who have been on this coffee programme are now using the approach for other crops/commodities.

The farmers are involved in training other farmers, for instance, the field school in Rakai (Jjongeza Coffee Farmer Field School) trained more than 500 out-grower farmers of Kaweri Coffee Company in CWD management and other good agricultural practises in 2006.

The creation of more FFS will greatly enhance the farmers' knowledge and give a chance for most farmers to get the much needed production information within relatively short time.

Coffee production campaign

Coffee stakeholders showed concern about the declining volume of coffee exports from Uganda. Consequently, a number of meetings/workshops were convened in Kampala during 2006, with the goal of improving coffee productivity, so as to raise the volume of coffee from 2 million bags to 4 million bags by the year 2015. A plan of action was drawn up for the various stakeholders, which included research, extension, input stockists and others. Control of CWD is essential if the programme is to succeed. Resistant varieties would greatly impact on the progress to increase production. The role of research is to ensure available resistant materials are evaluated, multiplied and distributed to farmers. At the same time, information dissemination through various print materials in English and main local languages on CWD management and other agronomic and crop protection problems is emphasized. The campaign began in 2006, and this is the third year.

3.10. Conclusion

Until resistant planting materials become widely available, farmers are strongly advised to continue to manage CWD through sanitary and cultural measures. These include regular inspection of their fields to identify trees with early symptoms, followed by uprooting and burning of the affected bushes *in situ*. Ensuring there is adequate soil fertility through addition of fertilizers or manure is also advocated. Preventing introduction of inoculum into the field or at least delaying introduction of inoculum into their fields through infected planting materials or contaminated soil is recommended, and thus, the source of planting materials should be from well-run, reputable nurseries where the soil is sterilized. Farmers should avoid use of volunteer seedlings from forests that could be harbouring the disease without showing symptoms. More than 40,000 farmers have already been trained and sensitized to the disease, and many are currently being trained in FFS in Rakai, Bugiri and Mukono.

The information gap that existed at the beginning of the CWD epidemic has currently been bridged. Participatory approaches to technology generation and dissemination have played a major role and are being adopted by extension workers. Local leaders have been sensitized about CWD and their pivotal role in the control of the disease. These leaders have realized that without coffee, their revenue base would remain small, and they are showing more interest in the crop than before, particularly in Masaka district and a number of others.

Recommendations for CWD management are now based on research findings. There are sound scientific reasons why uprooting and maintaining other sanitary practises are essential and are being communicated to farmers and extension staff. It is essential that information dissemination should be maintained to integrate new information from research into management strategies and keep farmers and extensionists updated.

Breeding work should continue to establish the nature of resistance in the current selected CWD-resistant cultivars to provide a basis for planning breeding strategies for durable resistance to CWD. Attempts to have DNA finger printing of the selected resistant clones for identifying them and for patent rights should continue to logical conclusions.

Acknowledgements

We gratefully acknowledge all the farmers and extension staff that participated in the programme with us; without their cooperation, patience, time and interest, no useful results would have been achieved. We would also like to thank UCDA for its commitment to this commodity and to all the development partners (CFC, ICO, EU, DFID, CIRAD etc.) for providing funds at the most needed time. In addition, we would like to acknowledge CORI staff for their dedication, in particular Mr Sammy Olal for his hard work in taking photographs and processing data.

References

Anon (1993) Background to the Budget for the Financial year 1993/1994. Ministry of Finance and Economic Planning. Uganda government.

Anon (1994) Background to the Budget for the Financial year 1994/1995. Ministry of Finance and Economic Planning. Uganda government.

Anon (2006) Background to the Budget for the Financial year 2004/2005. Ministry of Finance and Economic Planning. Uganda government.

Banana Bacterial Wilt Task Force Report (2002) Report for the Eradication of Banana Bacterial Wilt Disease in Kayunga and Mukona Districts. National Agricultural Research Organisation, Uganda, pp. 20–21.

Banana Bacterial Wilt Task Force Report, (2003) The National Research and Development Strategy and Action Plan for Control of Banana Bacterial Wilt. National Agricultural Research Organisation (NARO) Uganda.

Booth, C. (1971) The Genus Fusarium. CMI, Kew, Surrey. Pp 28, 107, 118–121.

Coste, R. (1992) Coffee. The Plant and the Product. Macmillan Press Lt., London, Pp 5–8.

Coffee Research Institute (CORI) Annual Report for 1996–1997.

CWD Task Force Report (1999) The National Research and Development Strategy and Action Plan for Control of Coffee Wilt Disease NARO, pp. 7–8.

De Graaf, J. (1986) The economics of coffee: economics of crops in developing countries, No 1. Pudoc, Wageningen, The Netherlands, pp. 21–26.

Flood, J. (1996) A study of the tracheomycosis or vascular wilt disease of coffee in Zaire (IMI). Report presented to Zairian Coffee Organisation (OZACAF). Pp 13.

Flood, J. (2005) Epidemiology and variability of *Gibberella xylarioides*, the coffee wilt pathogen. Final Technical Report. CABI Bioscience, p. 32.

Gerlach, W. and Nirenberg, H. (1982) The genus *Fusarium* – a pictorial Atlas, Berlin, Pp 7–9, 297–300.

Hakiza, G.J., Kyetere, D.T. and Olal, S (2004) Mode of penetration and symptom expression in Robusta coffee seedlings inoculated with *Gibberella xylarioides*, the cause of coffee wilt disease in Uganda. *Proceedings of the 20th International Conference on Coffee Science*. Bangalore, India. pp. 1231–1234.

Jameson, J.D. (Ed) 1970 *Agriculture in Uganda*. Oxford University Press, London.

Ladela, A.A. (2001) Beyond training and visit: a sustainable extension approach for Africa through a phased participatory extension education system. *African Crop Science Conference Proceedings* 5, 805–810.

Leakey, C.L.A. (1970) The improvement of Robusta coffee in East Africa. In: Leakey, C.L.A. (Ed) *Crop Improvement in East Africa*. CAB. pp. 250–277.

Lukwago, G. and Birikunzira, J.B. (1997) Coffee wilt disease (tracheomycosis) and its implications on Uganda's economy. *African Crop Science Conference Proceedings*. Pretoria. pp 969–974.

Marseden, D. (1994) Indigenous management and improvement of indigenous knowledge. In: Scoones and Thompson (Eds) *Beyond Farmer First*. Intermediate Technology Publications. London. pp. 52–56.

Muller, R. (1997) Some aspects of past studies conducted in Western and Central Francophonw Africa on Tracheomycosis (Cote d'Ivoire, Cameroon and Central African Republic). *Proceedings of the first Regional Workshop on Coffee Wilt Disease (Tracheomycosis)*. International Conference Centre, Kampala, Uganda. pp. 15–26.

Purseglove, J.W. (1968) *Tropical Crops. Dicotyledons 2*. Longmans Green and Co. Ltd London, Harlow, pp. 458–491.

Thomas, A.S. (1935) Types of coffee and their selection in Uganda. *Empire Journal of Experimental Agriculture* 12, 1–12.

Thomas, A.S. (1947) The cultivation and selection of Robusta coffee in Uganda. *Empire Journal of Experimental Agriculture* 15, 65–81.

Uganda Coffee Development Authority (UCDA). Annual Reports for 1999.

Uganda Coffee Development Authority (UCDA). Annual Report for 2004.

Uganda Coffee Development Authority (UCDA). Annual report for 2005.

Uganda Coffee Development Authority (UCDA). Annual report for 2006.

Wrigley, G. (1988) *Coffee. Tropical Agriculture Series*. Longmans, London.

4 Coffee Wilt Disease in Ethiopia

A. Girma,[1] A. Million,[1] H. Hindorf,[2] Z. Arega,[1]
D. Teferi[1] and C. Jefuka[1]

[1]*Jimma Agricultural Research Center, Ethiopian Institute of Agricultural Research, PO Box 192, Jimma, Ethiopia*
e-mail: girma.adugna@yahoo.com
[2]*INRES-Phytomedizin, University of Bonn, Nussallee 9, 53115 Bonn, Germany*

4.1. Introduction

Unlike Carl Linnaeus, contemporary scholars and writers about coffee authentically confirm that the primary centre of origin and genetic diversity of *Coffea arabica* L. (arabica coffee) is Ethiopia, and some of them wish to rename this species as *Coffea abyssinica*. Wild forest coffee is still found today in the south-western parts of the country. The intact afromontane rainforests of that region are centres of biodiversity of flora and fauna. Arabica coffee has been disseminated from this original place to many other countries like Brazil, Colombia and Vietnam, which are the leading producers and suppliers of coffee beans to the world market. Coffee has rapidly become one of the prominent commodity crops in global transactions, and it stands first in earning foreign currency for many countries including Ethiopia.

Ethiopia has the longest tradition of coffee production and consumption in the world with a traditional way of cultivation and the performance of inimitable 'coffee ceremony'. The country is also the oldest exporter of the world's finest original arabica coffees, some of which are unique and world-renowned speciality coffees, such as Harrar, Jimma, Limu, Sidamo, Yirgacheffe, Ghimbi and Lekempt, which receive premium prices. Coffee is crucial to the Ethiopian economy because it contributes 10% of the country's gross domestic product and generates more than 40% foreign exchange earnings.

Coffee remains crucial to the biological, social and economical values of the country, but despite being the birthplace of coffee, Ethiopia has not exploited and benefited from the crop to the best of its genetic and ecological potential. Coffee production systems remain predominantly traditional, and diseases and insect pests greatly reduce the productivity and quality of the produce.

This chapter reviews the status of coffee wilt disease (CWD), including its occurrence, distribution and importance on arabica coffee in Ethiopia, and highlights some of the factors accelerating the disease and efforts made to contain the problem.

4.2. The Importance of Coffee in Ethiopia

Ethiopia stands first in Africa and sixth in the world with regard to production of arabica coffee, and substantial increases have been achieved from 245,650 t in 2000 to 320,000 t in 2007 (Fig. 4.1). More than 90% of the production is from the garden, semi-forest and forest coffee systems of small-scale farmers, whereas the remaining 10% comes from large-scale plantation coffee. Coffee is by far the number one export crop for Ethiopia, contributing about 40% of the country's foreign currency income. The total export revenue from coffee was previously much higher (ranging from 60% to 70%), but since 2000, the proportion of exchange earnings of the crop has declined to around 40%. This decline is due to substantial price increases and increased production of other export commodities such as oil crops, pulses, leather and leather products, fruits and flowers for the international market (Kassahun and Getnet, 2008). Coffee is directly or indirectly a source of livelihood for more than about 25 million people engaged in production, processing and marketing of the crop.

In addition, the potential economic value of the Ethiopian wild coffee gene pool as a resource for international coffee breeding programmes has

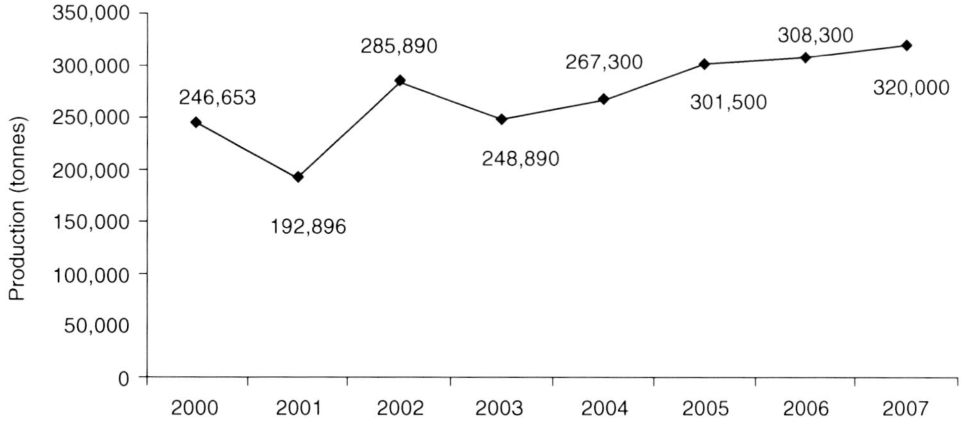

Fig. 4.1. Coffee production status over the past 7 years in Ethiopia (2000–2007) (Kassahun and Getnet, 2008).

been estimated to amount to *c.* US$0.5 and US$1.5 billion per year (Hein and Gatzweiler, 2006). This genetic resource provided a unique opportunity to develop more than 15 coffee berry disease (CBD)-resistant cultivars through rigorous research within a 5-year period. The immediate provision of seeds from these resistant materials to coffee farmers in the early 1970s saved the country's economy from catastrophe. Coffee yields have steadily increased over the last four decades – from 260 kg/ha in the early 1960s to more than 900 kg/ha during the late 1990s. Total Ethiopian production has risen to more than 3.8 million bags (over 2 million bags for export) with a stable overall quality; the amount of wet processed 'mild' coffee has increased since liberalization. Admasu (2006) estimated that a compound value amounting to US$270,317,647 (2297.7 million birr) was gained in the period 1972–2001 due to the increase in coffee yield by planting the improved cultivars on about 20% of land area.

4.3. Constraints to Coffee Production

Although there has been a considerable increase in coffee yields and overall production, the average coffee productivity in Ethiopia remains low (about 500 kg/ha per year) as compared to the world standard and to other coffee-producing countries, namely, Brazil, Vietnam, India and Kenya. Coffee is largely produced by small-scale subsistence farmers and is a low input–output crop in Ethiopia. It grows under age-old traditional production systems (forest, semi-forest and garden). Agronomic practices and crop husbandry remain conventional, and the adoption and diffusion of improved technologies, e.g. high-yielding cultivars and better pre- and postharvest management activities, have been slow. Coffee diseases such as CBD, CWD and coffee leaf rust (CLR) cause severe crop losses. Coffee insect pests, mainly antestia, leaf miners and coffee berry borer, cause damage in most coffee-growing regions of the country. Volatility and fluctuation in coffee prices, be it at local or international markets, can have enormous impacts to farmers' livelihoods and to investment decisions in the coffee industry as a whole.

4.4. Most Significant Disease Problems and Pests

Fifteen diseases (Eshetu *et al.*, 2000) and more than 45 insect pests (Million, 2000) have been recorded on *C. arabica* in Ethiopia, and among others, CBD, CWD and antestia are foremost in reducing the quantity and quality of coffee in the country. CBD (*Colletotrichum kahawae*) can cause up to 100% yield loss, with national average losses varying between 25% and 30% (Van der Graaff, 1983; Merdassa, 1986; Eshetu *et al.*, 2000). CWD is prevalent in almost all coffee-growing regions, with national average incidence and severity of 28% and 5%, respectively. CLR (*Hemileia vastatrix*) occurs in most coffee areas with varying intensities but is more prominent on Harrar coffee, and as high as 27% severity was reported (Eshetu *et al.*, 2000). Antestia (*Antestiopsis intricata*) is the major coffee pest inflicting considerable damage, amounting 9% berry fall and 48% darkened coffee beans (Million, 2000).

4.5. Brief History of CWD in Ethiopia

Historically, CWD on *C. arabica* was first observed in Ethiopia (Keffa province) by Stewart (1957), who described the wilting symptom and also identified the causal organism to be *Fusarium oxysporum* f.sp. *coffeae*. Lejeune (1958) also noted the presence of this disease on arabica coffee. Later, based on comparative studies of the isolates collected from dying arabica coffee trees from different origins and different *Coffea* spp., the causal was confirmed to be *Gibberella xylarioides* Heim & Saccas, of which *Fusarium xylarioides* Steyaert is the imperfect (conidial) state (Kranz and Mogk, 1973). Van der Graaff and Pieters (1978) reported that this pathogen caused a typical vascular wilt disease and was the main factor of coffee tree death in Ethiopia.

Subsequent surveys accompanied by isolation and identification demonstrated occurrence of *G. xylarioides* (*F. xylarioides*) in major coffee-growing regions of south and south-west Ethiopia (Van der Graaff and Pieters, 1978; Merdassa, 1986; Girma, 1997; Eshetu *et al.*, 2000). Even in some localities like Bebeka and Teppi, CWD outbreaks were noticed in large-scale plantation coffee (Girma, 1997, Eshetu *et al.*, 2000). During recent years, the prevalence and importance of CWD have been markedly increasing throughout coffee-producing areas of the country (Girma *et al.*, 2001; Girma and Hindorf, 2001; CABI, 2003; Girma, 2004; Oduor *et al.*, 2005).

CWD (also known as tracheomycosis) kills the whole coffee tree within a short period. Infected arabica coffee plants (of any age) usually occur singly or at random in groups in the affected fields. The earliest symptom of infection on both mature and young coffee trees is epinasty of leaves on some branches in the lower tree canopy. The leaves become brown or dark brown within two or more weeks and finally drop off the branches. These external symptoms most frequently start on one side of an infected coffee tree (unilateral or partial wilting) but eventually advance so that the whole plant is affected. Later in the season, completely wilted trees become desiccated and severely defoliated (Plate 4). These trees cannot be easily pushed over and uprooted as opposed to coffee trees that have died from root rot disease (*Armillaria mellea*). Internally, brown or blue–black discoloured bands are seen on the exposed wood of the stem (Plate 8). Dark fruiting bodies (perithecia) of the pathogen can be observed on the bark of stems especially around the collar region (Plate 7) and occasionally on branches of dead coffee trees (Girma *et al.*, 2001). Similar symptom description was given by Van der Graaff and Pieters (1978) and Flood (1997).

The characteristic partial wilting symptom accompanied by discoloured internal tissues effectively facilitates diagnosis and recognition of infected coffee trees in the field. This early detection allows rouging out of the infected trees early in the season before fungal sporulation at the advanced stage of pathogenesis (Girma and Hindorf, 2001).

4.6. Coffee Wilt Distribution in Ethiopia

Coffee production in Ethiopia is broadly grouped into four systems on the basis of biological diversity of the species and level of management, namely, forest, semi-forest, garden and plantation coffee (Meyer, 1965; Paulos and Demel, 2000). Forest coffee (sometimes referred to as 'wild coffee') regenerates spontaneously from self-sown seedlings as in the understorey of intact multilayered tropical rainforests. The system is characterized by a very rich genetic diversity of both coffee and other flora and fauna. This coffee-growing system is situated in the west and the south-west of the country.

Semi-forest or semi-domesticated coffee is simply derived from the wild forest coffee through human intervention and domestication, i.e. by thinning the dense 'overstorey' of the forest trees and slashing the understorey bushes and shrubs. The open areas are filled mostly by transplanting naturally propagated coffee seedlings from under the mother trees resulting in irregularly spaced trees with a high population density. Slashing is practiced once a year just before or during coffee picking season. Semi-forest coffee also contains diverse populations of coffee and contributes 35% to the total coffee production of the country, although its productivity is low (Paulos and Demil, 2000).

The garden coffee system is predominant in the southern and south-eastern regions, with plots of varying size (usually less than 0.5 ha) around farmers' dwellings and predominantly intercropped with a variety of fruit, root and cereal crops. The coffee population is less diverse consisting of many landraces. This system involves more intensive management practices such as slashing and hoeing (three to five times per year), and to some extent, the coffee trees are pruned, mulched and fertilized with organic materials (Workafes and Kassu, 2000).

Plantation coffee is a relatively new farming system for coffee cultivation in Ethiopia. The system involves monoculture of plants derived from CBD-resistant and high-yielding cultivars in nurseries and transplanting them into well-prepared land. Improved agronomic practices such as row planting, correct spacing, intercropping, mulching, pruning, shade regulation and to some extent advanced postharvest technologies are being applied. Higher yield and good-quality coffee are obtained from plantation coffee.

CWD occurs in all of the above coffee production systems (Fig. 4.2) to varying extent of damage among and within coffee fields and districts (weredas) depending on different interacting factors, mainly susceptibility of coffee trees, intensity of cultural practices and environmental conditions (Merdassa, 1986; Girma and Hindorf, 2001; CABI, 2003; Girma, 2004).

4.6.1. CWD in the forest and semi-forest coffee

The occurrence of CWD was reported after assessment in four forest coffee areas in south-west and south-east afromontane rainforests with incidence

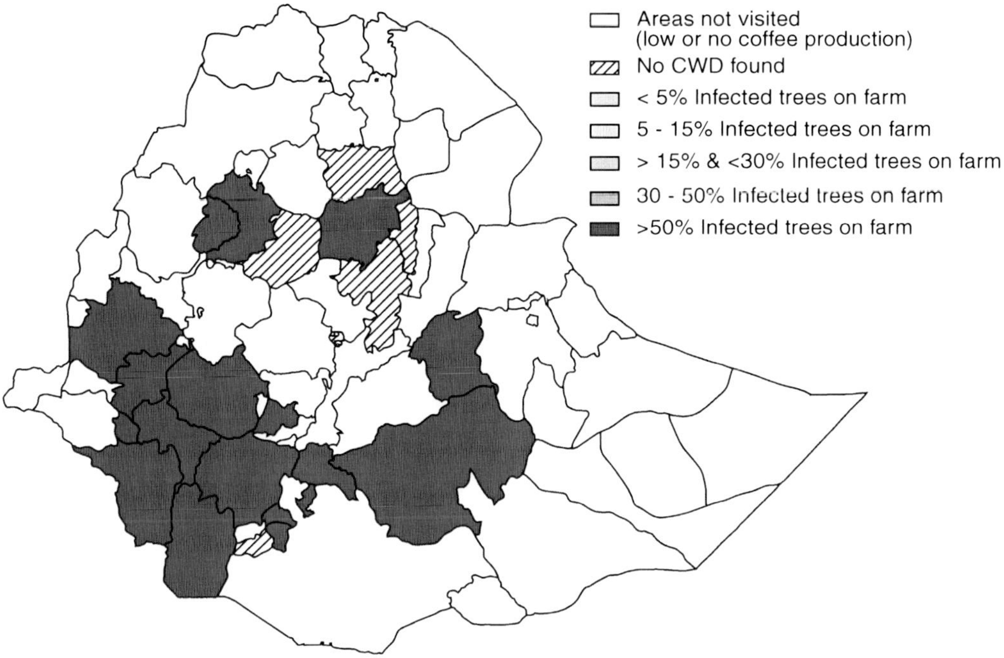

Fig. 4.2. Average severity (% trees infected per farm) of CWD in different zones in Ethiopia (CABI, 2003).

ranging between 5% at Sheko and 30% at Yayu. Although it was indicated that the damage was minimal in the dense stands of coffee (Van der Graaff, 1983; Merdassa, 1986), this was the first documented report that showed presence of CWD on forest coffee trees. Arega (2006) also demonstrated increasing occurrence of CWD in some forest areas like in Harenna (Bale) and Bonga (Keffa). The mean incidence in semi-forest coffee ranged from 3.6% at Mettu to 15.5% at Gera situated in south-west coffee-producing areas (Fig. 4.3), and the severity varied between 18.6% and 25.4% in some coffee fields at Yirgacheffe (Fig. 4.4) (Girma, 2004). A similar situation was observed in Bale, Jimma, Ilubabor, and West Wellega zones (CABI, 2003).

4.6.2. CWD in garden coffee

CWD is prevalent in the southern region, specifically in the three major quality-coffee-producing districts, namely, Wonago, Kochore and Yirgacheffe of Sidama and Gedeo zones, with highest incidence in Yirgacheffe followed by Kochore and Wonago. The severity of wilting in the sample fields in Yirgacheffe varied between 27.2% and 43.5% in the garden coffee as compared to that of the semi-forest coffee (Girma, 2004) (Fig. 4.4). Although the disease was not evenly distributed in most coffee-growing areas of Southern

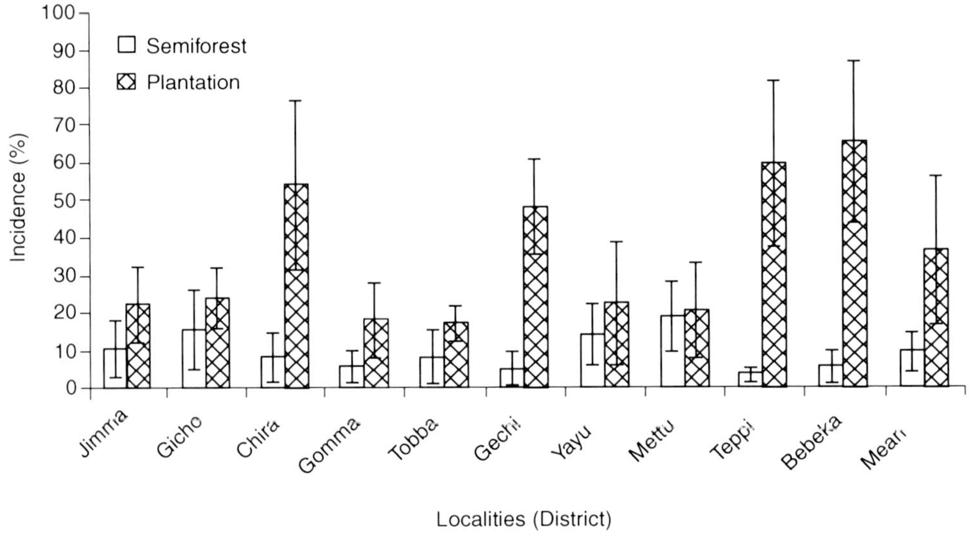

Fig. 4.3. Incidence of CWD in the semi-forest and plantation coffee production systems under farmers' conditions in south-western Ethiopia (error bars indicate standard deviations).

Nation, Nationalities and Peoples region, the average incidence (35%) and severity (5.0%) was significantly ($P < 0.001$) higher than in other regions. It was particularly high in the Sidama and Gedeo zones, with an incidence over 90% and severity of 25% (Fig. 4.5). The incidence of CWD was also above 35% in garden coffee of West Gojam zone of Amhara regional state, but it was very low in Wolaita (Southern Nation, Nationalities and Peoples) and West Harerghe (Oromiya) (CABI, 2003).

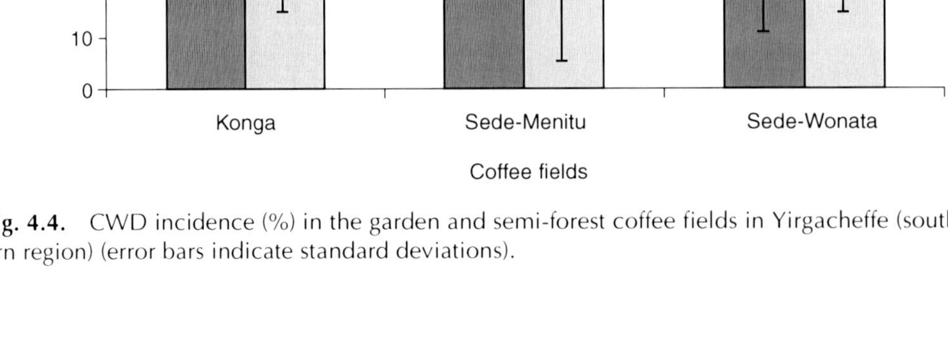

Fig. 4.4. CWD incidence (%) in the garden and semi-forest coffee fields in Yirgacheffe (southern region) (error bars indicate standard deviations).

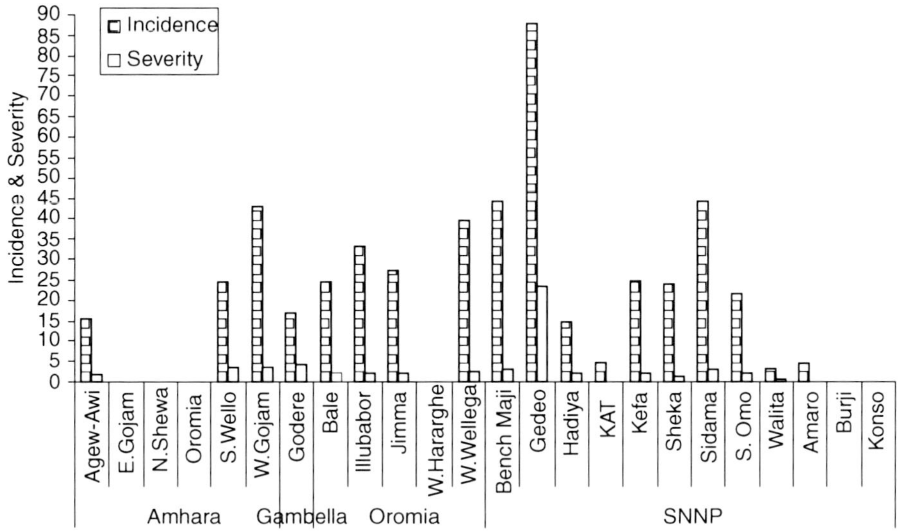

Fig. 4.5. Incidence and severity (%) of CWD in different regions and zones of Ethiopia in 2002 (CABI, 2003).

4.6.3. CWD in plantation coffee

The disease incidence is more severe in plantation coffee such as at research centres, on larger farmer holdings (1 to 5 ha) and in large estate commercial farms. CWD is commonly encountered in the research plots at Gera and Jimma amounting 42.5% and 48.2%, respectively (Table 4.1). It is serious in the farmers' coffee plantations at the Gera, Chira and Gechi districts, with respective mean incidence ranging from 21.7% to 25.5%, from 32.3% to 77% and from 35% to 60%, respectively (Table 4.2). The overall mean coffee tree loss in the farmers' plantation was more than 30%, and in total, about 10 ha of plan-

Table 4.1. Prevalence and incidence of CWD in various experimental plots of coffee research centres in Ethiopia.

Research centres/stations	Number of fields (*n*)	Incidence (%) Range	Incidence (%) Mean and SD[a]	Altitude (m)
Jimma	10	19.8–82.0	48.2 ± 23.1	1750
Agaro	3	5.2–12.1	8.7 ± 3.4	1650
Gera	15	21.0–61.1	42.5 ± 18.7	2000
Mettu	3	23.3–30.9	27.1 ± 5.4	1550
Teppi	3	6.5–13.4	10.0 ± 4.9	1200
Wenago	3	5.7–14.6	9.8 ± 4.5	1850
Mean		5.2–82.0	24.4 ± 17.7	

[a]SD = standard deviations.

Table 4.2. Incidence (%) of CWD plantation coffee under farmers' condition in south-west Ethiopia.

Location	Field	Estimated area (ha)	Incidence (%)	
			Range	Mean
Gera	Gicho 1	1.0	11.5–35.0	24.5
	Gicho 2	1.5	8.7–38.0	21.7
	Sedi-Loya	1.0	23.9–27.1	25.5
Chira	Gure-Genji	5.2	38.0–75.0	51.5
	Chira 1	4.5	55.0–89.0	77.0
	Chira 2	1.5	14.0–42.0	32.3
Tobba	Yachi	0.3	12.1–20.8	16.5
	Kilole	0.4	14.6–23.9	19.3
	Ageyu	0.2	8.3–27.0	16.1
Gomma	Shashamene	0.5	12.7–19.4	10.8
	Echemo	0.3	12.5–15.5	13.6
	Sombo	0.2	25.8–34.2	29.2
Gechi	Camp	0.5	25.0–70.0	48.9
	Mine-kobba	5.0	15.0–55.0	35.0
	Asendabo	5.0	37.7–78.6	59.7
Yayo	Jitto	1.0	11.0–34.0	22.5
Mettu	Sor	0.5	8.0–33.3	20.4
Mean	Total = 17	Total = 28.6 ha	8.3–89.0	30.9 ± 18.2

tation coffee had been abandoned completely. The lowest percentage of the disease was recorded in the farmers' plantation at Tobba (17.3%), whereas the highest was at Bebeka (65.2%) (Fig. 4.3). Girma *et al.* (2001) confirmed that the disease was more severe in plantation coffee at Bebeka, Teppi, Gera and Jimma. Van der Graaff (1979) remarked that some spectacular failures of the modern plantations system could be due to *G. xylarioides*, and when comparisons are made across production systems, the disease is more destructive in garden and plantation coffees than in forest and semi-forest coffee systems. The latter two systems are composed of heterogeneous coffee populations possessing varying levels of resistance and less human interference. However, in the former systems, characterized by relatively homogenous coffee trees and high levels of intervention, the disease spreads from tree to tree, from row to row and from one block to the other developing throughout the field (Girma, 2004). A remarkable increase in CWD severity of (11.5%) was recorded over a 6-month period in nine districts (weredas) of Gedeo and Sidama zones of Ethiopia (CABI, 2003).

4.7. Importance of CWD in Ethiopia

For many decades, CWD was considered as a minor problem in Ethiopia, and its impact therefore largely remained unnoticed and its effect underestimated, but the losses incurred due to the disease are comparable to those caused by CBD. With CWD, the whole tree dies and all neighbouring coffee trees die, so there is a loss of capital to the farmer; CBD affects only cherries. In addition, CBD can be controlled relatively easily with fungicides, but CWD is a soil-borne pathogen and this presents difficulties in the application of chemical treatments; affected fields may need to be left as fallow for some years or other crops planted (Girma, 2004).

Coffee production (yield) at the farm level decreased by 37% (from 1482 to 932 kg per sample farm), and this led to a decline in income of 67% (from 5038 to 1651 birr). The annual national crop losses attributed to CWD was 3360 t amounting to US$3,750,976 in Ethiopia (CABI, 2003). This economic loss coupled with difficulty to manage the disease indicates that CWD is the second leading disease of coffee, after CBD in Ethiopia.

4.8. Factors Influencing Severity of CWD in Ethiopia

Some preliminary observations indicate that temperature, rainfall, topography, coffee tree age, shade, soil type and weeding methods have significant effects on CWD. The incidence of CWD is higher on coffee trees that are older, shaded, planted on loamy soil and weeded by slashing (CABI, 2003). Above all, susceptibility of coffee cultivars and agronomic/cultural practices have consistent effect on the occurrence and severity of the disease.

4.8.1. Susceptibility of coffee cultivars in the field

A number of researchers have reported existence of marked differences in resistance levels in arabica coffee populations to CWD under field conditions at various locations (Van der Graaff and Pieters, 1978; Merdassa, 1986; Girma and Hindorf, 2001; Girma, 2004). Merdassa (1986) assessed the incidence of the disease in single-tree progenies of different coffee accessions for 6 years (1979–1984) at Gera and obtained tree loss ranging from 0.3% to 87%.

In a field at Bebeka, 23 cultivars (including four introduced Catimor lines) were planted in a completely randomized block design with three replications and 90 trees per plot. Cultivar 785, 1185, 1785 and 4485 were uniformly attacked in all plots and showed significantly high mean death rates of 80.0%, 72.9%, 83.4%, and 97.4%, respectively, indicating their susceptibility to coffee wilt. In contrast, the Catimor lines (1579, 1779, 1979 and 2179) and some French collections (F-15, F-27 and F-59) had the lowest infection levels of less than 10% (Girma, 2004). The introduced coffee lines such as Caturra Rojo, Caturra Amerello and Catuai showed significantly ($P < 0.05$) higher mean incidences of 83.0%, 80.5% and 80.0%, respectively, and were more susceptible to CWD than the indigenous cultivars 7454, 74110, 74112, 74140 and 74165 at Teppi (Girma, 2004). In all cases, the disease developed dramatically into large foci starting from a single tree, and spread is faster in plots composed of susceptible trees.

4.8.2. Agronomic/cultural practices

Besides the inherent genetic factors of coffee cultivars, certain agronomic and routine cultural practices have long been postulated to aggravate CWD. There is a strong association with wounding and dissemination of the fungal fruiting bodies (perithecia and ascospores) from a single infected tree to other disease-free trees/plots. The predominant disease-spread mechanisms mainly involve human activities including pruning, stumping (to rejuvenate old and unproductive trees), slashing and hoeing to control weeds and transporting infected trees from one field to the other (long-distance spread). The use of contaminated farm implements across various plots/fields also plays a significant role in spreading the fungus inoculum. A common practice in Ethiopia is to cut wilted trees, store them somewhere in the field or near the houses and use for various purposes such as for firewood, for fencing around dwelling houses or coffee farms and as a support for climbing beans. These trees and remaining stumps harbour the fungal fruiting bodies (perithecia with ascospores) that serve as inoculum source for further infection and initiate disease epidemics (Girma, 2004). It has been estimated that 60% of the farmers in Ethiopia used the wood for fencing, 26% for constructing houses and animal sheds, 10% gave surplus wilted trees to their neighbours for firewood and 2% sold the trees (CABI, 2003).

4.8.3. Spacing and population density of coffee trees

The severity of CWD seems to be affected by spacing between coffee trees and population densities. The results of disease assessments in a population density trial at Gera, consisting of eight cultivars planted in a circular fashion 'fan design' with each circle as a spacing treatment, showed that the mean incidence of disease linearly increased and significantly rose from 16.0% for 2 × 2 m to 28.5% for 1 × 1 m spacing between trees (Table 4.3). The disease outbreak

Table 4.3. Severities (%) of CWD in arabica coffee population density trial at Gera.

Coffee cultivars	Spacing between coffee trees (m)					
	2.0 × 2.0	1.75 × 1.75	1.5 × 1.5	1.25 × 1.25	1.0 × 1.0	Mean[a]
741	29.2	22.2	41.6	44.4	47.2	36.9 B
744	12.5	18.1	25.0	15.3	34.7	21.1 A
7440	13.9	9.7	18.1	22.2	20.9	16.9 A
7454	30.6	37.5	50.0	45.8	38.9	40.6 B
74148	8.3	5.6	12.5	9.7	20.8	11.4 A
74158	19.4	12.5	11.1	16.7	26.4	17.2 A
74165	4.2	4.2	15.3	11.1	8.4	8.6 A
75227	9.7	2.8	9.7	22.2	30.6	15.0 A
Mean[a]	16.0 a	14.1 a	22.9 ab	23.4 ab	28.5 b	

[a]Means followed with the same letter(s) are not significantly different, least significant difference values ($P = 0.05$) for cultivars and spacing treatments are 10.9 and 8.6, respectively.

started mostly at the centre of the plot where trees were planted very close to each other (1 × 1 m) and radiated outwards as spacing distances between trees increased (2 × 2 m). This fact may be ascribed to root-to-root contact via which the fungus can be transmitted from infected to nearby healthy trees. The lateral and feeder roots of arabica and robusta coffees can spread in the soil for a distance of 1.2 to 1.8 m from the trunk (Wrigley, 1988). There is ample evidence that *G. xylarioides* is abundantly recovered from root parts of symptomatic and asymptomatic trees (Girma *et al.*, 2001).

In addition, closely spaced trees are more liable to wounding, and cross inoculation from diseased to healthy trees can occur during slashing or hoeing coffee fields. Almost all stems of coffee trees have wounds at the crown level or a few centimetres above, and on average, 98% of the diseased and 95% of healthy trees were noted to have one to three wounds per coffee stem (Girma, 2004). The wounds arose practically from the intensive slashing of weeds in coffee fields by machetes, which is the most common method of weed control in coffee in Ethiopia (Getachew, 1991; Tadesse, 2001). Getachew (1991) noted that weeds are slashed frequently, sometimes more than ten times a year, depending on the dominating weed flora in plantation coffee, and most of the coffee trees were found to have at least one wound.

The effect of wounding was further demonstrated in young coffee seedlings in the greenhouse. When seedlings with healthy roots were transplanted into either naturally or artificially infested soils, no wilting symptoms appeared. Infections were exhibited when the tap roots were injured and transplanted into artificially or naturally infested soils. In addition, only those seedlings inoculated by stem wounding through nicking with *G. xylarioides*-infested scalpels or by injecting conidial suspensions into the stems became infected. The stem-nicking inoculation method also illustrates the role of contaminated farm implements in cross-inoculating coffee trees and disseminating the pathogen across coffee fields (Girma, 2004).

4.8.4. Replanting into the fields infested with the pathogen

Replanting susceptible coffee cultivars into already contaminated fields increases fungal inoculum, and further infection occurs in the second planting, e.g. at Gera, where about 57.8% and 36.8% infection were recorded during the first and second plantings in the same field (Girma, 2004; Girma *et al.*, 2007). Similarly, Girma (1997) estimated 40.0% tree death on young coffee replanted as 'refill' on sites of previously uprooted infected trees of the same coffee line at Teppi. Among some socio-economic factors noted contributing to the spread of CWD, particularly in Ethiopia, was frequent replacing/replanting with several seedlings (three to eight) per uprooted wilted trees (CABI, 2003). The infection of the young replants undoubtedly suggests that the fungus survives in stump and other debris and the soil for some 2 years after uprooting.

Perithecia of *G. xylarioides*, containing a great number of viable ascospores with a germination rate of 90% to 95%, are very abundant in the coffee fields. These sexual structures are the most important sources of inoculum in

CWD epidemics. High infection of susceptible arabica seedlings was noticed after inoculating with field-collected ascospores, suggesting that the perithecial state is the primary source of inoculum in the field (Girma, 2004).

4.9. Country-Specific Management Practices in Ethiopia

4.9.1. Cultural control

Unlike with other coffee diseases, namely, CBD and CLR, coffee trees infected by CWD cannot be saved. Successful control of the disease depends on the principles of disease prevention (avoid wounding of any part of the plant) and phytosanitation. The conventional phytosanitary approach of uprooting and burning the whole infected coffee tree on the spot is strongly recommended to coffee farmers to contain the disease as soon as symptoms are seen, but this relies on early diagnosis. Use of CWD-infected trees for any purpose is prohibited, and replanting with susceptible coffee seedlings should be delayed at least for 2 years (Girma *et al.*, 2001; Girma, 2004). Cultural weed control activities like slashing and digging should be avoided in CWD-prone coffee fields, and agronomic practices (pruning and stumping) that bring about wounding in coffee trees should be done with efficiently disinfected tools. Disinfection of farm implements such as machetes, bow saws and pruning shears with potent disinfectants (>75% alcohol) followed by intense heating with fire is strongly recommended to farmers whenever pruning, rejuvenating old coffee trees and thinning newly suckers. Farmers' field schools recommend growing cover crops such as *Desmodium* sp. and haricot bean, which are very efficient in suppressing weeds (so reducing the need for slashing) and as legumes, promote the growth of coffee trees. Applying ash, mulch and slashing between plots with hand weeding around coffee trees were also promising treatments in CWD control trials in Ethiopia (Fig. 4.6) (CABI, 2005, unpublished).

4.9.2. Chemical control

Preliminary results of on-station and on-farm participatory trials implicate that coffee stem paint with copper oxychloride (Kocide) and weed control with herbicide reduced CWD incidence (Fig. 4.5) (CABI, 2005, unpublished). Fungicide paint of the cut surface of stumped coffee with Kocide is observed to effectively protect the large wounds from infection by the CWD pathogen.

4.9.3. Biological control

The result of a recent *in vitro* study conducted by Muleta *et al.* (2007) on antagonistic effects of some rhizobacteria against the three *Fusarium* pathogens including *F. oxysporum*, *Fusarium stilboides* and *F. xylarioides* were promising. Of 23 bacterial isolates obtained from rhizospheres of arabica coffee trees

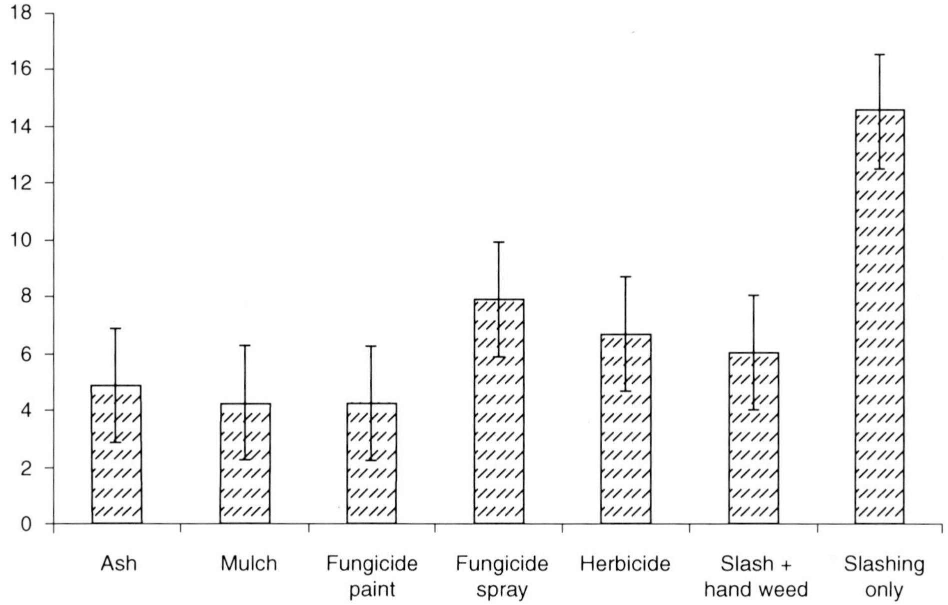

Fig. 4.6. Effect of different agronomic practices and control methods on CWD in Ethiopia (error bars indicate standard deviations) (CABI, unpublished).

in south-west Ethiopia, 21 significantly inhibited the mycelial spread of *F. xylarioides*. *Bacillus subtilis*, designated as isolate 'AUBB20', was the most antagonistic to this pathogen and was indicated as a potential candidate for biological control of CWD (Muleta *et al.*, 2007).

4.9.4. Deployment of CWD-resistant cultivars

In fields affected by CWD, it is possible to observe considerable variation in disease severity among coffee cultivars (Van der Graaff and Pieters, 1978; Merdassa, 1986; Girma, 1997; Girma and Hindorf, 2001; Girma *et al.*, 2001; Girma, 2004). Cultivar SN-5, F-51/53 and 248/71 showed 100% tree loss, whereas F-35 and F-51 had significantly ($P < 0.05$) lower mortality rates of 9.3% and 27.9%, respectively, at Gera (Girma, 1997; Girma *et al.*, 2001). At the same locality in another highly infested field where some CBD-resistant selections ($n = 30$) were planted, disease incidence ranged from 12% for selection 8150 to 96% for 74304 (susceptible check) (Girma, 1997; Girma and Hindorf, 2001).

Apparently, disease-free (tolerant) trees in the field can be selected and evaluated for true resistance through repeated seedling inoculation tests under controlled conditions (Chapter 10). Artificial inoculation tests have shown that cultivars 1579, 200/71 and 8136 were resistant to CWD with low-percentage deaths (12.7%, 15.2% and 25.2%, respectively) accompanied by long incubation periods (Table 4.4) before symptoms appeared (Girma and Chala, 2008). Cultivars 146/71, 206/71 and 8144 showed moderate CWD infection, whereas others including Caturra and Geisha had the highest wilt severity (>90%) indicating susceptibility to the disease. There was a correlation between the lowest seedling death rates in the greenhouse and wilt severity observed in the fields (Table 4.4). Thus, those cultivars demonstrating resistant reactions under both field and greenhouse conditions can be recommended for use in CWD-prone areas provided that they have other desirable traits like resistance to CBD, high yield and improved quality.

4.9.5. Training and information dissemination on CWD management

Because a large number of coffee farmers (>85%) were not aware of CWD prior to the Regional CWD Program (CABI, 2003), the first step was training

Table 4.4. Reactions of some coffee cultivars to CWD.

Coffee cultivars	Actual value (mean % death)	Transformed value[a]	Incubation period (mean no. of days)	Incidence in the field[b]
1185	86.0	75.1 ab	90.0 op	75.0
1785	78.7	67.9 a–h	80.0 p	75.0
1579	12.7	16.9 s	157.5 a	10.2
2179	63.3	53.4 i–o	140.8 a–d	20.5
4/70	77.2	62.0 b–l	117.5 d–m	56.2
36/70	60.9	56. f–n	92.5 n–p	15.3
146/71	34.6	35.1 qr	122.5 d–k	68.4
200/71	15.2	20.3 s	152.5 afb	28.2
206/71	52.8	46.1 m–q	125.0 d–j	48.9
8112	74.9	63.2 a–k	112.5 f–o	63.1
8133	64.2	54.1 g–n	122.5 d–k	19.2
8136	25.3	29.6 rs	150.0 a–c	29.4
8143	61.6	52.7 j–p	125.0 d–j	42.4
8144	40.2	39.1 o–r	137.5 a–e	37.0
F-27	81.0	67.0 a–j	90.0 op	10.9
F-35	85.7	70.9 a–e	97.5 l–p	26.2
Caturra	68.9	59.2 d–m	130.0 b–h	74.5
Geisha	88.1	73.9 a–c	97.5 l–p	29.1
7440[c]	40.4	38.7 p–r	135.0 a–f	20.3
SN-5[d]	69.7	56.7 e–n	119.2 d–l	99.9
Mean	68.8	58.3	115.5	–
LSD[e] ($P < 0.05$)		14.4	23.5	–
CV (%)		21.8	17.9	–

[a]The actual data (2 years) were transformed to arcsine square root values before analysis. Means followed with the same letter(s) are not significantly different from each other.
[b]CWD incidences (mean %) summarized from various fields and localities (Girma, 1997; Girma and Hindorf, 2001; Girma, 2004).
[c]Resistant/tolerant cultivar.
[d]Susceptible checks.
[e]LSD = least significant difference.

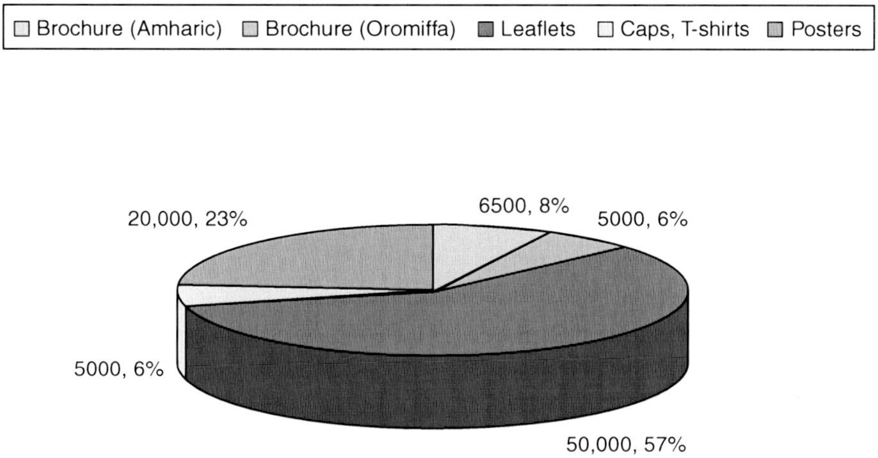

Fig. 4.7. Proportions of CWD publicity materials produced and disseminated in Ethiopia.

extension agents who in turn trained coffee farmers about the disease. The training courses included diagnosis and identification of CWD-infected trees based on symptoms and signs of the pathogen, transmission/spread mechanisms and control methods. Besides, a large number different types of publicity materials such as leaflets, brochures, manuals, posters, wall calendars, T-shirts, caps and stickers with brief CWD information were produced and disseminated to almost all coffee stakeholders in the country. Exercise books describing CWD were provided to school students in the vicinity of most affected areas (Fig. 4.7). There have been publications in newspapers and bulletins and broadcasts on television and radio (national and local FM) notifying farmers of CWD prevention and control. Efforts were made to raise awareness and sensitize policy/decision makers at various levels via conferences and national workshops. Participatory on-station and on-farm trials and season-long farmers' field schools on CWD management practices were conducted, and farmers' field days were organized to popularize promising results of the trials to coffee farmers in highly affected districts (weredas) in south and south-west Ethiopia (see Chapter 11).

4.10. Conclusion

The primary foci of CWD in Ethiopia are not known because the disease was reported to occur concurrently in many areas of the country. In general, it appears that CWD existed for many years as an endemic disease of *C. arabica* but has gained importance over time in almost all coffee-growing regions. The nationwide biological survey of CWD showed that on average, 27.9% of 1607 sample coffee farms were affected, with disease incidence ranging from 15% to 34.0% and disease severities varying between 1.3% and 5.0% (CABI, 2003;

Oduor *et al.*, 2005). However, it has been realized that CWD incidence was relatively higher and more severe in the garden-based and plantation farming systems than in the semi-forest and forest coffee production systems. Among other factors, varietal composition and human activities profoundly aggravate the distribution and significance of the disease in Ethiopia.

The soil-borne nature of the pathogen and perennial character of coffee have made management of the disease difficult through the conventional control approach of 'uproot and burn infected trees at the spot'. Avoidance of using infected trees for firewood, for construction of huts and fences or for other agricultural uses and avoidance of immediate replanting/replacing with susceptible coffee seedlings have been recommended in Ethiopia (Girma *et al.*, 2001; Girma, 2004). Any practice that create wounds in coffee trees like slashing and digging of coffee fields should be avoided, whereas hand weeding around the trees, spraying herbicides, application of mulch and growing cover crops such as *Desmodium* sp. are encouraged. All farm implements need to be efficiently disinfected for use in coffee fields. Knowledge transfer and information dissemination especially descriptions of symptoms have enhanced the early diagnosis of this disease and have greatly contributed to containing the disease. In fact, in the short term, sensitization of major coffee stakeholder of the threat of CWD remains one of the key practices in managing this disease because early diagnosis can aid management.

The longer-term prospects of successful management of CWD depend principally upon employing resistant coffee cultivars. In this regard, the genetic variability of arabica coffee populations presents a great opportunity to develop a number of CWD-resistant varieties. Some cultivars, such as 1579, 200/71 and 8136, have shown high-resistance levels in artificial seedling inoculation tests that well correlated with that of natural infection in the field. To exploit the enormous genetic potential in the control of CWD, independent selection and screening programme should be designed and implemented as experienced in the ever successful CBD programs.

In addition, from a research perspective, biocontrol agents including *Rhizobacteria*, *Trichoderma* spp. and other potential microorganisms should be tested against CWD pathogen as for other *Fusarium* spp.

Acknowledgements

We appreciate CABI for the excellent coordination and execution of the Regional Coffee Wilt Programme and the European Union and the Common Fund for Commodities for funding and fully supporting implementation of the programme. The German Academic Exchange Service is also acknowledged for financial support of the research work conducted on CWD in Ethiopia. This work would not have been possible without the commitment and determination of the staff of Jimma Agricultural Research Centre in particular and the Ethiopian Institute of Agricultural Research in general. Dr J. Flood (CABI, UK) is indeed highly appreciated for critically reading and improving this chapter.

References

Admasu, S. (2006) Economic analysis of conserving wild coffee genetic resource in southwestern Ethiopia. PhD thesis. University of Bonn, Bonn, Germany.

Arega, Z. (2006) Diversity of arabica coffee populations in afromontane rainforests of Ethiopia in relation to *Colletotrichum kahawae* and *Gibberella xylarioides*. MSc thesis. School of Graduate Studies, Department of Biology, Addis Ababa University, Addis Ababa, Ethiopia.

CABI (CAB International) (2003) Surveys to assess the extent and impact of coffee wilt disease in East and Central Africa. Final technical report. CABI Regional Centre, Nairobi, Kenya. pp. 149.

Eshetu, D., Teame, G. and Girma, A. (2000) Significance of minor diseases of *Coffea arabica* L. in Ethiopia: a review. In: *Proceedings of the workshop on control of coffee berry disease (CBD) in Ethiopia.* Addis Ababa, Ethiopia, pp. 58–65.

Flood, J. 1997. Tracheomycosis or vascular wilt disease of coffee in Uganda. Report presented to Ugandan Coffee Development Authority (UCDA), Uganda. 12 pp.

Getachew, Z. (1991) A study of coffee weed flora and possible control measures in coffee state farms of Kaffa, southwestern Ethiopia. MSc thesis. Alemaya University of Agriculture, Alemaya, Ethiopia.

Girma, A. (1997) Status and economic importance of *Fusarium* wilt disease of arabica coffee in Ethiopia. In: Hakiza, G.J., Birkunzira, B. and Musoli, P. (eds.) *Proceedings of the First Regional Workshop on Coffee Wilt Disease* (*tracheomycosis*). International Conference Centre, Kampala, Uganda, pp. 53–61.

Girma, A. (2004) Diversity in pathogenicity and genetics of *Gibberella xylarioides* (*Fusarium xylarioides*) populations and resistance of *Coffea* spp. in Ethiopia. PhD dissertation. University of Bonn, Bonn, Germany.

Girma, A., Flood, J., Hindorf, H., Bieysse, D., Simons, S. and Mike, R. (2007) Tracheomycosis (*Gibberella xylarioides*) – a menace to world coffee production: evidenced by cross inoculation of historical and current strains of the pathogen. In: *Proceedings of the 21st International Scientific Conference on Coffee Science* (*ASIC*). Montpellier, France, pp. 1268–1276.

Girma, A. and Hindorf, H. (2001) Recent investigation on coffee tracheomycosis, *Gibberella xylarioides* (*Fusarium xylarioides*) in Ethiopia. In: *Proceedings of the 19th International Scientific Conference on Coffee Science* (*ASIC*). Trieste, Italy, pp. 1246–1252.

Girma, A. and Chala, J. (2008) Resistance levels of arabica coffee cultivars to coffee berry disease, coffee wilt and leaf rust diseases in Ethiopia. In: *Proceedings of the 12th Conference of the Crop Science Society of Ethiopia* (*CSSE*). Addis Ababa, Ethiopia. *Sebil* 12, 92–103.

Girma, A., Mengistu, H. and Hindorf, H. (2001) Incidence of tracheomycosis, *Gibberella xylarioides* (*Fusarium xylarioides*) on arabica coffee in Ethiopia. *Journal of Plant Diseases and Protection* 108, 136–142.

Hein, L. and Gatzweiler, F. (2006) The economic value of coffee (*Coffea arabica*) genetic resources. *Ecological Economics* 60, 176–185.

Kassahun, H. and Getnet, G. (2008) Performance, structure and prospects of coffee marketing in Ethiopia. In: Adugna, G., Bellachew, B., Shimber, T., Taye, E. and Kufa, T. (eds.) *Coffee Diversity and Knowledge: Proceedings of a national workshop four decades of coffee research and development in Ethiopia.* Addis Ababa, Ethiopia, pp. 416–423.

Kranz, J. and Mogk, M. (1973) *Gibberella xylarioides* Heim & Saccas on arabica coffee in Ethiopia. *Phytopathology Z.* 78, 365–366.

Lejeune, P. 1958. Rapport au Gouvernement Impérial d'Ethiopie sur la production caféière. FAO, Rome. 49 pp.

Merdassa, E. (1986) A review of coffee diseases and their control in Ethiopia. In: Abate, T. (ed.) *Proceedings of the First Ethiopian Crop Protection Symposium.* Institute of Agricultural Research, Addis Ababa, Ethiopia, pp. 187–195.

Meyer, F.G. (1965) Notes on wild *Coffea arabica* from southwestern Ethiopia, with some historical considerations. *Economic Botany* 19, 136–151.

Million, A. (2000) Significance of arthropod pests of coffee in Ethiopia. In: *Proceedings of the Workshop on the Control of Coffee Berry Disease (CBD) in Ethiopia, Addis Ababa.* Institute of Agricultural Research, Addis Ababa, Ethiopia, pp. 66–71.

Muleta, D., Assefa, F. and Granhall, U. (2007) In vitro antagonism of Rhizobacteria isolated from *Coffea arabica* L. against emerging fungal coffee pathogens. *Engineering* in *Life Sciences* 7, 1–11.

Oduor, G., Phiri, N., Hakiza, G., Million, A., Asiimwe, T., Kilambo, D.L., Kalonji-Mbuyi, A., Pinard, F., Simons, S., Nyasse, S. and Kebe, I. (2005) Surveys to establish the spread of coffee wilt disease, *Fusarium* (*Gibberella*) *xylarioides,* in Africa. In: *Proceedings of the 20th International Scientific Conference on Coffee Science* (*ASIC*). Bangalore, India, pp. 1252–1255.

Paulos, D. and Demel, T. (2000) The need for forest germplasm conservation in Ethiopia and its significance in the control of coffee diseases. In: *Proceedings of the Workshop on Control of Coffee Berry Disease* (*CBD*) *in Ethiopia.* Addis Ababa, Ethiopia, pp. 125–135.

Stewart, R.B. (1957) Some plant diseases occurring in Kaffa province, Ethiopia. Imperial Ethiopian College of Agriculture and Mechanical Arts, Alemaya, Ethiopia.

Tadesse, E. (2001) Weed flora and weed control practices in Ethiopia: a review. In: *19th International Scientific Colloquium on Coffee (ASIC).* Trieste, Italy.

Van der Graaff, N.A. and Pieters, R. (1978) Resistance levels in *Coffea arabica* L. to *Gibberella xylarioides* and distribution pattern of the disease. *Netherlands Journal of Plant Pathology.* 84, 117–120.

Van der Graaff, N.A. (1979) Breeding for stable resistance in tropical crops: strategies to maintain balanced pathosystems in modern agriculture. *FAO Plant Protection Bulletin* 27, 1–6.

Van der Graaff, N.A. (1983) Durable resistance in perennial crops. In: Lamberti, L., Waller, J.M. and van der Graaff, N.A. (eds.) *Durable resistance in crops.* Plenum Press, New York, pp. 263–276.

Workafes, W. and Kassu, K. (2000) Coffee production systems in Ethiopia. In: *Proceedings of the Workshop on Control of Coffee Berry Disease* (*CBD*) *in Ethiopia.* Addis Ababa, Ethiopia, pp. 99–107.

Wrigley, G. (1988) Coffee. Tropical agriculture series. Longman Scientific and Technical Publisher, New York.

5 Status of Coffee Wilt Disease in Tanzania

D.L. Kilambo, N.M. Ng'homa, J.M. Teri and L. Masumbuko
Tanzania Coffee Research Institute (TaCRI), Lyamungu PO Box 3004, Moshi, Tanzania
www.tacri.org

5.1. Introduction

Coffee species grown in Tanzania are *Coffea arabica* and *Coffea canephora*. Arabica coffee is grown in the northern and southern highlands, Tarime and the west (Kigoma), whereas robusta coffee is grown in the western zone, mainly in the Kagera region. The total area under coffee in Tanzania is estimated to be 265,343 ha (Tanzania Coffee Board, 2006), which is composed of 77% arabica coffee and 23% robusta coffee. Most of the coffee farms produce around 250 kg of clean coffee per hectare.

Since its appearance in Tanzania in 1997, coffee wilt disease (CWD) has clearly demonstrated both its ability to spread rapidly to new areas and cause serious losses. From 1997 to date, 44 of 100 wards in the Bukoba, Muleba, Misenyi and Karagwe districts are reported to be affected by the disease, causing monetary losses of $197,551. In Tanzania, control of the disease has involved uprooting and burning of the affected trees *in situ*, with particular emphasis on early diagnosis of the disease to reduce the risk of spread to other trees and farms.

This chapter describes the history of the disease in Tanzania, its geographical distribution, economic importance and control measures.

5.2. Importance of Coffee in Tanzania

Coffee is currently Tanzania's second most important export among traditional exports accounting for 23% of the countries total foreign exchange of US$267.1 (Fig. 5.1). Coffee contributed about US$61.4 million to export earnings (Tanzania Coffee Board, 2006). The coffee industry in Tanzania provides employment to 420,000 families, of which 90,000 are from the robusta-growing

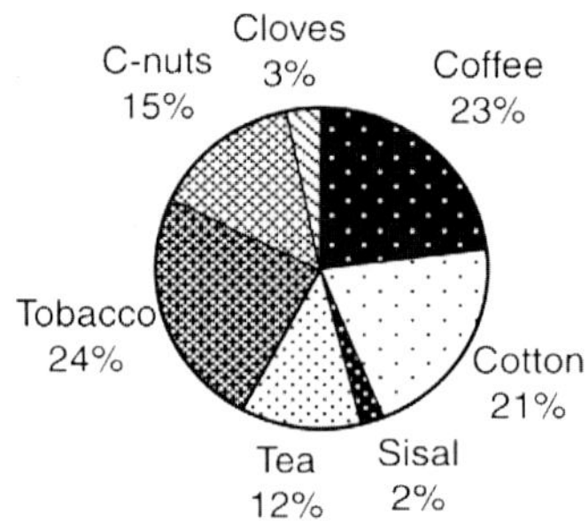

Fig. 5.1. Contribution proportional of Tanzania Traditional exports in 2006. (From the Tanzania Coffee Board.)

areas of the Kagera region, and benefits about 2,000,000 other people indirectly (Baffes, 2003).

5.3. Constraints to Coffee Production

There are a number of constraints that face coffee farmers in Tanzania. Productivity tends to be low with most of the coffee farms producing around 250 kg of clean coffee per hectare. Some of the factors that contribute to low coffee productivity include fluctuating world coffee prices and low farm gate prices resulting in small returns to coffee producers. In contrast to these low incomes, the costs of inputs such as fertilizers, fungicides and insecticides are extremely high, making them unaffordable to most of the farmers. Yet, these inputs are needed since the crop is threatened by a range of insect pests and diseases. In addition, due to a lack of information, many farmers use inappropriate husbandry practices.

5.4. Most Important Pests and Disease Problems

The production of arabica coffee in Tanzania is to a great extent limited by diseases and pests. These include coffee berry disease (*Colletotrichum kahawae* Waller and Bridge sp. nov.), coffee leaf rust (CLR; *Hemileia vastatrix* Berk and Br) and in some areas *Fusarium* bark disease (*Fusarium stilboides*). Among the insect pests, white stem borer (*Monochamus leuconotus*), antestia bug (*Antestiopsis lineaticollis*) and coffee berry borer (*Hypothenemus hampei* Ferrari) are the most serious pests of arabica coffee in Tanzania.

For robusta, coffee production is influenced by disease problems such as red blister disease on the berries caused by *Cercospora coffeicola* Cooke and Berk and CLR (*H. vastatrix* Berk and Br), which may cause serious leaf defoliation and dieback. The major insect-pest is coffee berry borer (*H. hampei* Ferrari), which if left unmanaged can cause yield losses of up to 90%.

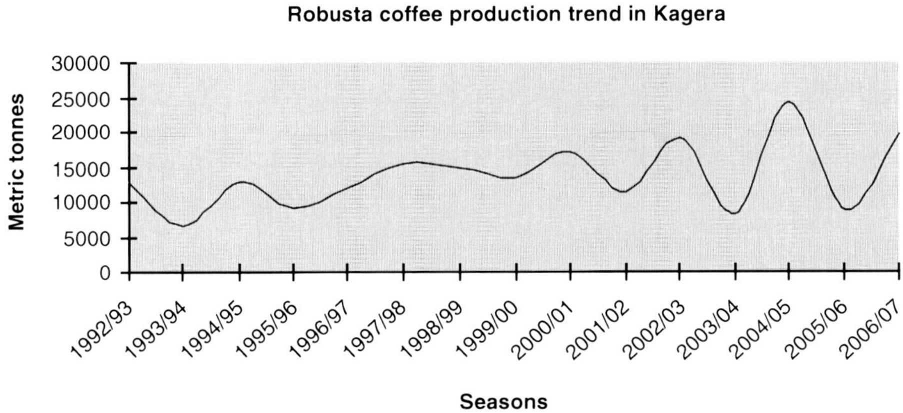

Fig. 5.2. Robusta coffee production trends in Kagera for 15 years. (From the Tanzania Coffee Board.)

More recently, robusta production in the Kagera region of Tanzania has been badly affected by a vascular wilt disease coffee (CWD). Since its appearance in 1997, coffee production trends in Kagera has been highly variable (Fig. 5.2).

5.5. Brief History of CWD in Tanzania

Incidences of a wilt-like disease in robusta coffee were first reported in Misenyi division in Minziro ward at Kigazi village near the Ugandan and Tanzanian border towards the end of 1996. A team of researchers and extension workers visited the affected areas in January 1997. A fungus, *Gibberrella xylarioides* Heim and Sacc, was isolated from robusta stems and branches collected from diseased plants (Kilambo and Kaiza, 1997; Kilambo *et al.*, 1997) and confirmed by CABI Europe (UK). Additional surveillance was made between 1998 and 2003 (Ng'homa, 2003; Mohamed *et al.*, 2002; Swai, 1998), and CWD was again confirmed in the Karagwe and Muleba districts. CWD has continued to be a serious threat to the robusta coffee industry in Kagera since this first report in 1997. Its introduction in Tanzania could be associated with human activities prevailing between residents of Kagera and neighbouring countries where there is CWD.

5.6. Coffee Wilt Distribution in Tanzania

A detailed biological survey was carried out in Tanzania between September 2002 and February 2003 to determine the geographical distribution and host range of CWD in Tanzania (Kilambo *et al.*, 2004; Oduor *et al.*, 2003). CWD was found to be confined only to robusta coffee in the Kagera region (Fig. 5.3).

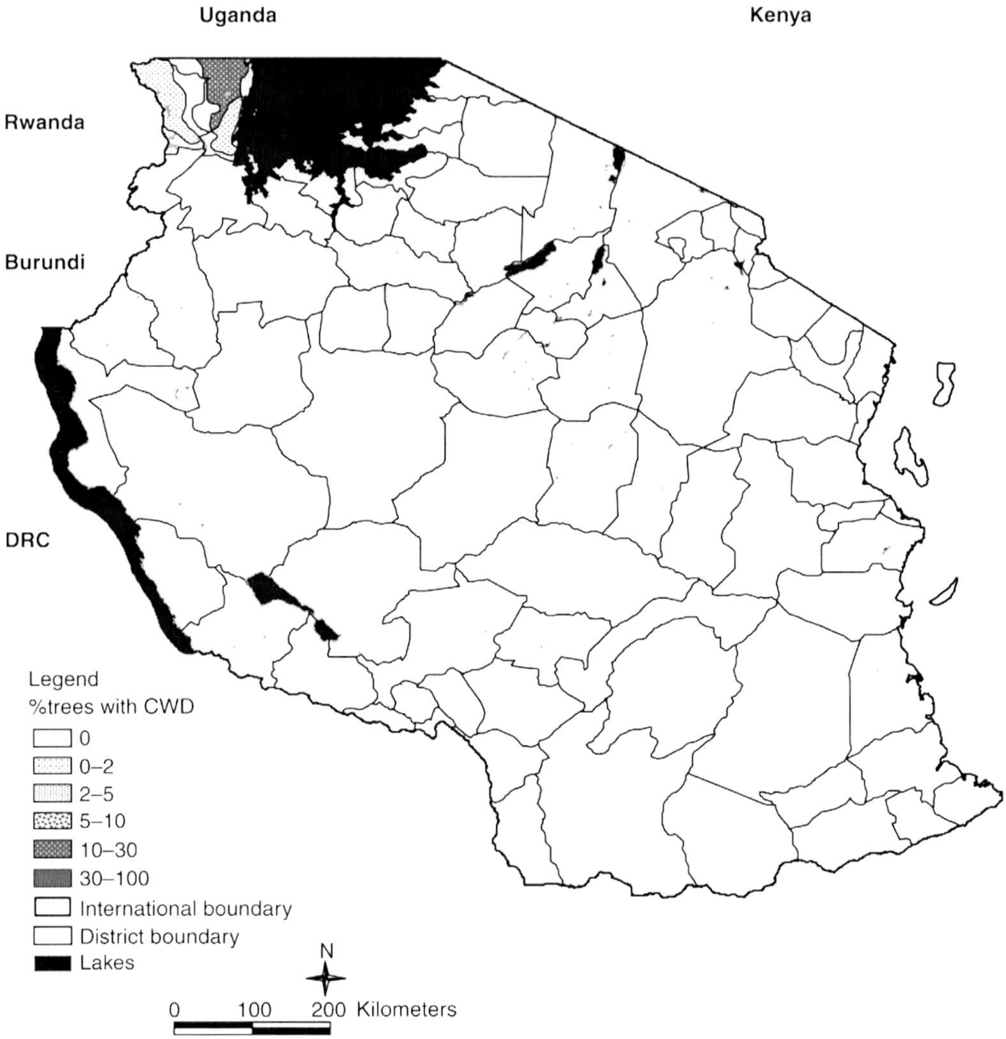

Fig. 5.3. Map showing distribution of CWD in Tanzania.

In the north part of Kagera where CWD is present, groups of infected farms are detected in close proximity to each other but about 100 km away, and in Muleba, only isolated farms were affected. This distribution could reflect the dispersal of the pathogen. Waller and Holderness (1997) indicated that spore dispersal (conidia and ascospores) can occur through different mechanisms, including rain, soil, plant-to-plant contact (relatively short distances), wind, humans and possibility of insects (for longer distance spread).

In Kagera, average incidence and severity were 13.5% and 4.6%, respectively, whereas higher incidences and severity were recorded in Bukoba (39.1% and 16.1%, respectively). Among divisions in Kagera, Bugabo was

Table 5.1. Estimated number of robusta trees infected and uprooted from 1997 to 2007. (From the Tanzania Coffee Research Institute Lyamungu, Moshi.)

District	Infected coffee trees	Uprooted coffee trees	Percentage of uprooted due to CWD
Bukoba	28,155	14,079	50
Misenyi	17,277	4,566	26
Karagwe	6,279	6,279	100[a]
Muleba	2,422	2,348	97
Total	54,133	27,272	50

[a]CWD is still prevailing.

the worst affected (incidences of 80.7% and severity of 46%). In 1997, CWD was detected in three wards, but to date (2008), the disease has spread and a total of 44 wards out of 100 in Kagera are now reported to be affected by the disease.

5.7. Importance of CWD in Tanzania

Prior to the outbreak of CWD, robusta coffee was the major source of income to more than 90% of households in Kagera. Since the outbreak of CWD, most of the coffee growers are in despair about the devastation caused by CWD. The disease has caused substantial losses of coffee trees in the region (Table 5.1). At the same time, cost of hiring labour for uprooting infected stems has increased from US$0.42 in 2002 to US$0.83 in 2007. Many farmers have switched livelihood as a coping strategy to the disease, with farmers shifting to fishing, growing bananas and brewing beer. In addition, many coffee farms have been abandoned. Losses of trees can be equated to yield loss,

Table 5.2. Estimated CWD infected trees, yield losses and cost of uprooting. (From Tanzania Coffee Research Institute Lyamungu, Moshi.)

District	Infected coffee trees 1997–2007	Estimated yield losses from CWD-infected trees[a] (kg)	Yield losses in monetary terms[+] (US$)	Cost of uprooting infected trees[#] (US$)	Total losses (US$)
Bukoba	28,155	84,465	141,057	23,369	164,426
Misenyi	17,277	51,831	86,558	14,339	100,897
Karagwe	6,279	18,837	31,458	5,212	36,670
Muleba	2,422	7,266	12,134	2,010	14,144
Total	54,133	162,399	271,207	44,930	316,137

[a]Estimated that a mature robusta coffee tree 25 years old can produce 3 kg of clean coffee.
[+]Estimated that the price of 1 kg of clean coffee is US$1.67 (as per October–December 2007).
[#]Cost of uprooting a stool of infected robusta tree is US$0.83.

which is approximately 162,399 kg of clean coffee lost due to the death of 54,133 trees from CWD. It is estimated that the disease has caused a financial loss of approximately US$316,137 over the last 10 years (Table 5.2).

5.8. Factors Affecting Severity or Incidence of CWD in Tanzania

The CWD epidemic in Tanzania can be mainly attributed to the following factors:

5.8.1. Improper gap filling procedures

New coffee seedlings are being planted immediately after uprooting of CWD-diseased trees. It is suspected that in the presence of *G. xylarioides* propagules, the roots of the new seedlings may start to be infected shortly after replanting.

5.8.2. Source of planting materials

Some farmers use volunteer seedlings that can either be already contaminated by *G. xylarioides* or be susceptible to the pathogen because they are from a narrow genetic base.

5.8.3. Use of infected material for fuel

CWD-infected robusta trees are being used as a source of firewood. Sometimes, this firewood is stored in the backyards of the farmers' houses. Following conducive weather conditions, the conidia of the asexual stage of the pathogen can be disseminated and induce infections of previously healthy coffee trees.

5.8.4. Quarantine procedures

Farmers are not following Plant Protection Act no. 13 of 1997, which prohibits movement of coffee planting materials and importing from neighbouring countries, from district to district and from farm to farm.

5.9. Country-Specific Management Practices in Tanzania

Since the outbreak of CWD, the government of Tanzania through the Ministry of Agriculture, in collaboration with other sectors of the coffee industry under the coordination of Tanzania Coffee Research Institute (TaCRI), has implemented strategies to minimize the effect of CWD in collaboration with regional partners (CABI Africa and Advanced Research Institutes in the region) and with international partners such as scientists based in the UK (CABI Europe, UK).

5.9.1. Exclusion

To restrict the spread of CWD, in 2002, the Ministry of Agriculture and Food Security by then enforced Plant Protection Act no. 13 of 1997 to effect quarantine measures to prevent movement of plant materials from neighbouring countries and also movement of plant material, soil and farm implements from district to district and from farm to farm. Achievements have been very low because enforcement was not very effective. Tanzania shares borders with other countries like Rwanda and Burundi, which do not yet have CWD, so there is a need to strengthen quarantine procedures to prevent spread of the disease to these countries also.

5.9.2. Eradication

During the past 10 years, joint efforts between TaCRI, CABI and internal stakeholders has greatly facilitated a reduction in the impact of CWD. In May 2003, stakeholders including smallholders, districts and regional leaders, agricultural extension staff, representatives from the Ministry of Agriculture and Food Security and Cooperative Unions met at Maruku Agricultural Research Institute and made a crucial decision that a campaign to eradicate CWD should be a top priority. A simple methodology was used, this involves the uprooting and burning of affected trees *in situ*, with particular emphasis on early diagnosis of the disease to reduce the risk of spread to other trees and farms. Results since the campaign have been impressive, with concerted awareness and education campaign helping to get the message to farmers effectively and rapidly. Eradication of CWD-affected trees has significantly reduced the incidence of the disease (Fig. 5.4). This shows that if eradication of CWD-diseased trees is done promptly in newly infected trees, it can assist in the management of the disease on the farm and minimize its spread to others. Since the outbreak of CWD, a total of 27,272 of diseased trees have been uprooted and destroyed. This is about 50% of infected coffee trees.

Coffee stakeholders supported efforts in locating areas affected by CWD, training programmes and eradication of affected trees. The support came from the Tanzania Coffee Board to initiate surveys in January 1997 by providing about US$8000; the Ministry of Agriculture and Food Security donated US$10,000 in 2002 to continue with the surveillance and eradication programme and districts councils in Karagwe and Muleba provided US$6000 for training of extension workers and for the eradication programme in 2002–2004. From 2002 to 2007, the Common Fund for Commodities donated over US$70,000 to conduct training of farmers and extension workers on the

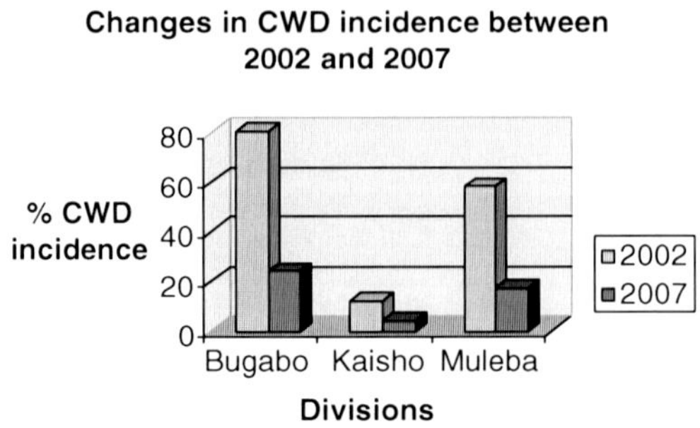

Fig. 5.4. Changes in incidence of CWD in three administrative divisions of Kagera between 2002 and 2007.

identification of early symptoms of CWD and dissemination of technologies to combat the disease. In addition, the Coffee Research Network (CORNET) supported by EU funds, contributed US$39,000 in 2002 to 2003 to carry out biological and socio-economic surveys. TaCRI (using STABEX funds) spent approximately US$10,000 in the eradication programmes.

5.9.3. Protection

Stem paint using copper oxychloride (300 g/l of water) applied 50 cm from ground level and four times a year as a preventive measure was shown to be effective in minimizing levels of CWD infection. This was tested in on-farm agronomic trials, and some farmers have started to adopt it. In addition, farmers' local practice of heaping ash at the collar position of robusta plants prevents infection of *G. xylarioides* in some of the areas, particularly in Bukoba. However, in Uganda, use of ash was not effective in preventing CWD in almost all agronomic trial sites. Efficiency of use in Tanzania may be associated with the soil pH of the particular areas where use of ash was effective. Observations in the Karatu district where lime is being applied in coffee farms have indicated that the treatment is effective for lowering the incidence of collar rot disease (*Fusarium lateritium*), but this requires further investigation. Measures to improve the health of trees by mulching, manuring and fertilizer application also assist the plants to tolerate CWD infection.

5.9.4. Host plant resistance

Commercial robusta clones MS 1, 2, 3, 4 and 5 released to coffee growers in recent years are all susceptible to CWD in varying degrees. Efforts to evaluate robusta coffee germplasm for their resistance to CWD using different isolates and the root dip technique developed by Hakiza *et al.* (2004) were initiated in May 2004. Seedlings (5–6 months old) were inoculated using the

Table 5.3. Eleven best-performing resistant robusta lines to CWD combining CLR resistance.

S/no	Coffee line/ cultivar	CWD resistance	CLR resistance
1	ML 26	R	R
2	NG 08	R	R
3	ML 35	R	R
4	NG 13	R	R
5	BK 27	R	MR
6	KR 21	R	MR
7	NG 12	R	MR
8	KR 11	R	MR
9	NG 17	R	MR
10	ON BK 02	R	MR
11	MR 10	R	MR
12	MS 1[a]	S	R

[a]Commercial cultivar, R = resistant, MR = moderately resistant, S = susceptible.

stem nicking method and the root dip method. In both methods, the inoculum was standardized at 1.3×10^6 conidia per millilitre of water. After 9 months of assessment, the survivors were considered as resistant and developed further for future selection (Table 5.3). To date, 273 clones have been established in a clonal mother garden at Maruku with the intention to obtain enough cuttings to establish in the National Performance Trials and on farms (Kilambo *et al.*, 2006).

5.9.5. Variability of the CWD pathogen

Twenty isolates were sent to CABI Europe (UK) for assessing genetic variability of the pathogen by studying the DNA banding patterns produced by PCR amplification of the ribosomal DNA intergenic spacer region of *G. xylarioides* isolates from coffee. The isolates were found to be genetically similar to larger *C. canephora* group from the Democratic Republic of the Congo and Uganda but different from that found to attack *C. arabica* in Ethiopia (Rutherford, 2005). Observation of CWD pathogenic variability in Kagera reveals that only *C. canephora* is being attacked by the pathogen.

Pathogenicity studies conducted in Tanzania showed that all 14 isolates collected from various ecosystems were very aggressive (Kilambo *et al.*, 2006) (Table 5.4). The isolates are currently being used to search for durable resistance.

A comparison of the methodologies for assessing resistance of coffee germplasm at Maruku was undertaken. Two methods for screening for

Table 5.4. Pathogenicity test results of *G. xylarioides* on MS 1 and MS 2.

CWD isolate code no.		Location collected			Number of dead seedlings at end of test	
TaCRI	CABI UK	District	Coordinates	Altitude (m)	MS 1	MS 2
2004/10	T 1	Muleba	S 01°45.901″; E 31°35.491″	1547	9	9
2004/13	T 2a	Muleba	S 01°46.827″; E 31°34.541″	1545	10	9
2004/07	T 3a	Muleba	S 01°49.702″; E 31°41.137″	1395	9	10
2004/08	T 4	Muleba	S 01°43.159″: E 31°38.078″	1510	10	9
2004/02	T 5a	Muleba	S 01°41.172″; E 31°37.731″	1287	10	9
2004/06	T 8a	Bukoba	S 01°00.595″; E 31°46.582″	1189	10	10
2004/01		Bukoba[a]	S 01°14.836″; E 31°50.682″	1200	10	9
2004/12	T 9a	Bukoba	S 01°01.612″; E 31°32.758″	1256	10	9
2004/14	T 12a	Karagwe	S 01°18.600″; E 30°47.205″	1424	9	9
2004/03	T 13a	Karagwe	S 01°26.166″; E 30°52.801″	1317	10	9
2004/05	T 14a	Karagwe	S 01°17.308″; E 30°53.896″	1659	9	9
2004/09	T 15a	Karagwe	S 01°15.309″; E 30°57.347″	1354	9	9
2004/09	T 15b	Karagwe	S 01°15.309″; E 30°57.347″	1354	9	9
2004/09	T 15c	Karagwe	S 01°15.309″; E 30°57.347″	1354	10	10
Mean					9.57	9.21
SE ±					0.14	0.11
CV					5.30	4.50
LSD[b] ($P \leq 0.05$)					0.30	0.23

[a]Isolate 2004/1 was used for CWD resistance evaluation.
[b]LSD = least significant difference.

resistance, namely, root dip and stem nicking, were evaluated for their ability to differentiate between susceptible and resistant material by producing a high selection pressure. In the study, the root dip method produced higher selection pressure by allowing fewer survivors than stem nicking (Kilambo *et al.*, 2006, Table 5.5).

5.9.6. International and regional collaboration

With support from DFID, two TaCRI scientists and a technician received in-depth training in appropriate pathogenicity methods and screening techniques at the Coffee Research Institute, Kituza, Uganda (Plate 32). This support also financed the collection and shipment of CWD isolates in UK and CWD screening for resistance. TaCRI is now applying these techniques for screening CWD resistance in this country. From 2004 to date, TaCRI used US$17,000 to support development of varieties with durable resistance to CWD and maintenance of robusta coffee germplasm.

Table 5.5. Robusta lines that survived after artificial inoculation with *G. xylarioides* using stem nicking and root dipping procedures.

S/no	Coffee line/ cultivar	Mean seedling survival at end of test	
		Stem nicking	Root dipping
1	ML 26	4	3
2	NG 08	5	3
3	ML 35	2	1
4	NG 13	4	3
5	BK 27	4	4
6	KR 21	3	3
7	NG 12	4	3
8	KR 11	5	3
9	NG 17	3	1
10	ON BK 02	3	3
11	MR 10	5	4
12	MS 1*	0	0
	Mean	3.50	2.58
	SE ±	0.40	0.35
	CV	40.00	35.40
	LSD ($P \leq 0.05$)	1.08	0.95

*Commercial variety.

5.9.7. Dissemination and training

As part of the programme, an awareness campaign was initiated which involved CWD training for 5659 growers and 192 extension staff in Kagera, and 171 in non-CWD areas. Training encompasses symptoms identification, safe handling and destroying of diseased robusta trees to minimize further spread. This was complemented by publication of articles in newspapers, radio and local television in Kagera Region and the publication of posters and leaflets alerting stakeholders (particularly farmers) of the threat from the disease. Approximately 1622 posters and leaflets have been distributed.

5.9.8. Farmers' field schools/participatory groups

Twenty-three groups have been formulated in CWD hot spot areas. The groups meet once a month, and farmers have an opportunity to share experiences in managing the effects of CWD. It is expected that knowledge of disease management will be disseminated to some other farmers. TaCRI has an innovative approach of participatory extension working with extension staff and farmer groups. This has assisted highly in the dissemination of technologies in minimizing the effect of CWD and boosting coffee production in the region. There are 23 participatory groups in Bukoba and Misenyi, 16 in Karagwe and 13 in Muleba. An excellent example of farmer-to-farmer extension is exemplified by Chabuhora, which had 30 participants and formulated five groups for training other farmers. This shows how a participatory approach through farmer-to-farmer contact, extension workers and researchers is effective in the diffusion of coffee husbandry practices. Manyafubu is another good example of a participatory farmer approach. The group have 25 members (13 men and 12 women) and was formed in March 2004. It conducted training to allow farmers to identify robusta trees infected with CWD and attempted to minimize spread through an eradication campaign. To date, the group has conducted training in 18 villages in the Bugabo division, and 1839 farmers including village leaders and ward representatives have been trained. Seven farmer field schools have been initiated; ten farmer groups formed in 20 training centres.

5.9.9. Open/field days

On 24 September 2004, TaCRI instituted an open day at Maruku. The major emphasis was to stress the threat of CWD, its impact on livelihoods and initial steps to consider on how to contain its further spread to new areas. More that 500 participants from all levels attended. Since then, 23 field days have been conducted in CWD-diseased areas in Kagera.

5.9.10. Sensitization of policy makers

Policy maker fora were used to disseminate information on CWD. For example, on 22 May 2003, TaCRI and representatives from the Ministry of Agriculture and Food Security were invited to attend the Regional Coordinating Committee of Kagera, chaired by the then Regional Commissioner Major Gen. Tumainiel Kiwelu. Highlights and strategies on management of CWD were presented. The meeting had significant impact on policy makers who supported the eradication programme and training of extension workers and farmers using funds from district councils.

5.9.11. Prioritizing CWD research programmes

In 2003, TaCRI formulated its 5-year Strategic Action Plan 2003–2008, whose vision is to contribute to the transformation of the Tanzanian Coffee Industry to sustainable prosperity with a major goal of creating a profitable and sustainable coffee industry in Tanzania. The vision of TaCRI is in line with the Agriculture Vision and the Tanzania Development Vision 2025. In this strategic action plan, CWD has been given a high priority. Since 2003, TaCRI has been addressing CWD by screening robusta germplasm present at Maruku and facilitating eradication and training programme at village levels.

5.9.12. Extensive replanting programme

Tanzania has embarked on an extensive coffee replanting programme with five carefully selected, indigenous robusta coffee clones being distributed to farmers, and from 1999, more than 1,500,000 coffee plantlets from a relatively narrow genetic base have been replanted by more than 1000 farmers. However, the coffee varieties used in this programme were later known to be susceptible to CWD, so there are renewed efforts to get new CWD-resistant clones to the farmers, as a replanting programme will have great impact in Kagera.

5.10. Conclusion

G. xylarioides is the most damaging pathogen of robusta coffee in Tanzania. Despite efforts already undertaken to minimize the effect of the disease, the answers related to the rational control measures will depend on successful execution of the following activities:

1. Concentration on already long-term breeding strategies for robusta coffee, involving:
   - Identification of more CWD-resistant accessions from Maruku germplasm by artificial inoculation.
   - Initiation of hybridization schemes and selection and advancement of elite materials.
   - Selection of local robusta 'survivors' from CWD hot spot areas.
   - Expansion of clonal mother garden already established at Maruku, hence providing more CWD-resistant materials for distribution.
2. Continue to disseminate information and train farmers about minimizing further spread and socio-economic impact of CWD.
3. Continue with national and international collaboration to have an in-depth knowledge of the CWD pathogen and to search for durable resistance through exchange of germplasm.

Acknowledgements

The management of CWD in Tanzania since its first report 10 years ago has been an excellent example of partnership and collaboration. We are grateful to the Tanzania coffee stakeholders, European Commission, Dar es Salaam and the Government of Tanzania for generous financial support to TaCRI during the implementation of this work. We are also grateful to DFID, CFC, CABI Africa and Europe-UK, the Coffee Research Centre, Uganda and Centre de Cooperation Internationale en Recherche Agronomique pour le Développement (CIRAD) for support to carry out activities in Kagera to combat the threat of CWD.

References

Baffes, J. (2003) Tanzania coffee sector: constraints and challenges in global environment. Africa Region Working Paper Series No. 56, The World Bank, Washington, DC.

Hakiza, G.J., Kyetere, D.T. and Olal, S. (2004) Mode of penetration and symptom expression in robusta coffee seedlings, inoculated with *Gibberella xylarioides*, the cause of coffee wilt disease in Uganda. In: *ASIC 2004 20th International Conference on Coffee Science*. Bangalore, India.

Kilambo, D. and Kaiza, D. (1997) Investigation of *Fusarium xylarioides* on robusta coffee in Misenyi division Bukoba district. Trip report, Lyamungu misc. report.

Kilambo, D., Kaiza, D. and Swai, F.B. (1997) Observations of wilt disease in some coffee growing areas in Tanzania. In: *Proceedings of the 1st Regional workshop on coffee wilt disease (tracheomycosis)*. International Conference Centre, Kampala, Uganda.

Kilambo, D.L., Ng'homa, N.M., Mohamed, R., Teri, J.M., Poole, J., Flori, A. and Pinard, F. (2004) Coffee disease surveys in Tanzania. In: *Proceedings of the 20th International Conference on Coffee Science*. Bangalore, India, pp. 1263–1266.

Kilambo, D.L., Ng'homa, N.M., Mtenga, D.J., Teri, J.M., Nzallawahe, T., Rutherford, M. and Masumbuko, L. (2006) Progress towards searching for durable resistance to *Fusarium* wilt (*Fusarium xylarioides*) in *Coffea canephora* germplasm on Tanzania. In: *Proceedings of the 21st International Conference on Coffee Science*. Montpellier, France, pp. 1386–1389.

Mohamed, R.A., Ng'homa, N.M., Sayi, B. and Kabumbire, A. (2002) *Report on coffee wilt disease Surveys in Kagera*. Ministry of Agriculture and Food Security, United Republic of Tanzania.

Ng'homa, N.M. (2003) Survey and mapping of areas affected by coffee wilt disease (CWD) in Karagwe district. Technical report, Ministry of Agriculture and Food Security, supported by Karagwe District Council.

Oduor, G., Simons, S., Phiri, N., Njuki, J., Poole, J., Pinard, F., Kyetere, D., Hakiza, G., Musoli, P., Lukwago, G. Abebe, M., Tesfaye, A., Kilambo, D., Asiimwe, T. and Munyankere, P. (2003) Surveys to assess the extent and impact of coffee wilt disease in East and Central Africa. Final Technical Report EU Contract No. ASA-RSP/CV-006. CAB International, Egham, UK.

Rutherford, M.A. (2005) Epidemiology and variability of *Gibberella xylarioides*, the coffee wilt pathogen. DFID – Crop Protection Programme, Final Technical Report CABI Uk. 24 pp.

Swai, F.B. (1998) Surveillance of wilt disease in Kagera. Lyamungu misc reports.

Tanzania Coffee Board (TCB) (2006) Data on coffee production. TCB miscellaneous reports.

Waller, J.M. and Holderness, M. (1997) Fusarium diseases on coffee. In: *Proceedings of the First Regional Workshop on the Coffee Wilt Disease (Tracheomycosis)*. International Conference Centre, Kampala, Uganda, pp. 31–39.

6 Socio-Economic Impact of Coffee Wilt Disease

R.O. Musebe,[1] J. Njuki,[1] S. Mdemu,[2] G. Lukwago,[3] A. Shibru[4] and T. Saiba[5]

[1]*CABI Africa, UN Avenue, ICRAF Complex, PO Box 633-00621, Nairobi, Kenya*
[2]*Tanzania Coffee Research Institute, Lyamungu, PO Box 3004, Moshi, Tanzania*
[3]*National Agricultural Research Organization, PO Box 421, Kabale, Uganda*
[4]*Jimma Agricultural Research Centre, Ethiopian Institute of Agricultural Research PO Box 192, Jimma, Ethiopia*
[5]*Office National du Café, BP 8931, Kinshasa Kingabwa, Congo*

6.1. Introduction

Coffee is a major export crop in Africa and accounts for the bulk of the export earnings of most economies in the Eastern and Central Africa (ECA) regions. It is also important because of its contribution to farm income, employment and food security. The crop is a major source of livelihood particularly for small-scale producers. Coffee is a leading export sector in Uganda's economy, accounting for over US$100 million in export sales in the 2000/2001 market year (GoU, 2001). In Kenya, coffee ranks fourth after horticulture, tea and tourism, accounting for 10% of the total export earnings (Karanja and Nyoro, 2002). Similarly, the commodity is Tanzania's largest export crop, contributing approximately US$115 million to export earnings and provides employment to over 400,000 families (Baffes, 2003). In Ethiopia, coffee contributes around 50% to the country's foreign exchange earnings and about 10% of the gross domestic product. More than 25% of the population of Ethiopia, which represents 15 million people, are dependent on coffee for their livelihoods, including 8 million people directly involved in coffee cultivation and 7 million in the processing, trading, transport and financial sectors (Beintema and Solomon, 2003; Oxfam, 2002; Charveriat, 2001). In the Democratic Republic of Congo (DRC), coffee is the number three export after copper and cobalt, and it represents approximately 9% of the total value of DRC exports and almost 60% to 65% of the total value for the whole of exports of the agricultural produce (Banque Centrale du Congo, 2005). Growth in coffee production and income has been one of the main engines for the development of coffee-producing areas. Thus, coffee plays a significant role in the national economies and the livelihood of the rural poor in the subregion.

Coffee production, productivity, quality and earnings are under threat from coffee wilt disease (CWD). Incidence and severity of CWD are highest in Uganda and lowest in Tanzania (Table 6.1). In Tanzania, the disease was only found in the Kagera region, which borders Uganda.

CWD is observed on robusta coffee only in DRC, Tanzania and Uganda; whereas in Ethiopia, it was observed on arabica coffee only. This finding suggests that the diseases found in the two coffee types, which are also separated geographically, are genetically distinct (Oduor *et al.*, 2003). Following the re-emergence of CWD, farmers have experienced changes in their livelihoods and resource endowments and have adopted various coping strategies as a result of the disease. This chapter examines some of the changes that have occurred in respect to coffee since the re-emergence of CWD as farmers try to deal with the problem in DRC, Ethiopia, Tanzania and Uganda and draws conclusions and recommendations aimed at improving coffee production through containment of the disease. The changes analysed include (i) the importance of coffee as a source of income, (ii) coffee production, (iii) input use in coffee, (iv) liquidation of assets, (v) coping strategies and (vi) household expenditure.

Different sampling approaches were used in different countries to identify the households to provide the requisite data. The sampling approach used depended on the distribution of CWD in the country, with purposive sampling being used where incidence of the disease was relatively low or focused in certain areas of the country and random sampling where CWD was widespread and severe. In Tanzania, CWD was found in the Kagera region only, and cluster sampling was used in villages known to have CWD, and these were identified through key informants; purposive sampling was used to target 99 affected farms. In Uganda, a multistage sampling procedure was followed in which 21 robusta-coffee-growing districts affected by CWD were targeted; subcounties, parishes and villages were subsequently selected randomly. Three farmers were then selected from each village. A total of 356 households were interviewed. In Ethiopia, the socio-economic survey was

Table 6.1. Incidence and severity of CWD in countries in the ECA region.

	Incidence (%)	Severity (%)
DRC	90.0	40.5
Ethiopia	27.9	3.0
Tanzania	2.2	0.7
Uganda	90.3	44.5

1

2a

2b

Plate 1. Coffee field attacked by CWD as shown by stumps of dead trees. (Photograph courtesy of G. Hakiza.)

Plate 2. (a) Healthy coffee field in 1993. (b) The same field in 1998. (Photograph courtesy of G. Hakiza.)

3

4a

4b

Plate 3. Robusta plant affected by CWD in Uganda. (Photograph courtesy of S. Olal.)
Plate 4. (a and b) Arabica coffee trees with complete wilt symptoms. (Photographs courtesy of A. Girma.)

5

6

7

8

Plate 5. Veinal necrosis, caused by coffee wilt disease (*Gibberella xylarioides*), clearly visible on the underside of a coffee leaf. (Photograph © M. Rutherford.)

Plate 6. Wounds (arrows), originating from slashing during weed control, at the base of a coffee stem. (Photograph courtesy of N. Phiri.)

Plate 7. Stromatic fruiting bodies (perithecia and ascospores of *Gibberella xylarioides* in the bark of dead Arabica coffee tree. (Photograph courtesy of A. Girma.)

Plate 8. Characteristic blue-black colouration under the coffee bark when an infected stem is scraped with a knife. (Photograph courtesy of N. Phiri.)

9

10

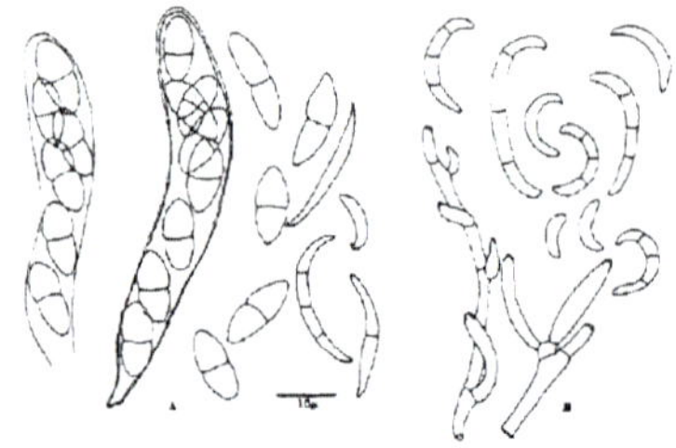

11

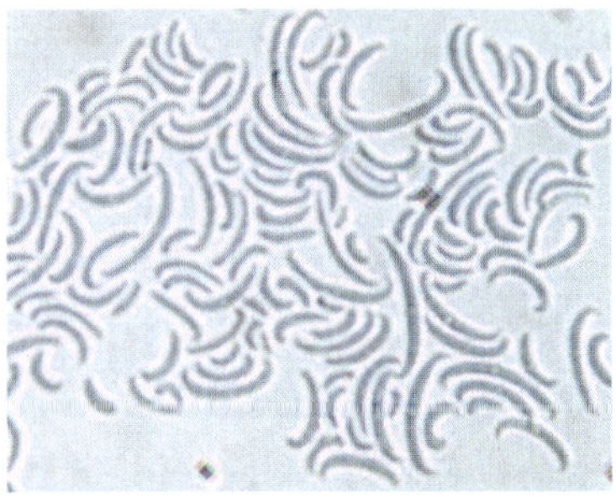

12

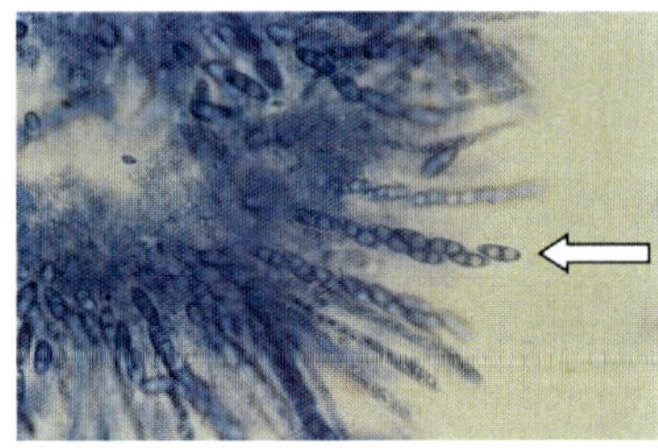

13

14

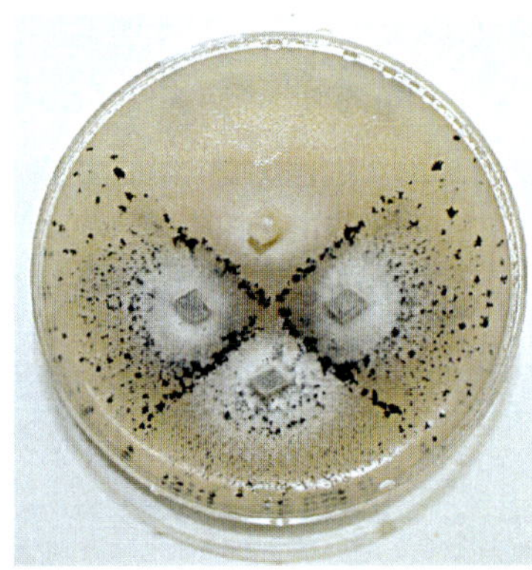

Plate 9. Perithecioide ascomata ('perithecia') produced by the coffee wilt pathogen, *Gibberella xylarioides,* in cracks of the tree bark. (Photograph © J. Flood.)

Plate 10. Diagrammatic representation of asci (left) and the typically curved conidia (right) of the coffee wilt pathogen, *Gibberella xylarioides.* Asci contain eight two-celled ascospores. Illustration from Booth, C. and Waterston, W.M. 1964. *CMI Descriptions of Plant pathogenic Fungi and Bacteria No. 24,* CABI, Wallingford.

Plate 11. Conidia of *Gibberella xylarioides* on Spezieller Nährstoffarmer Agar (SNA) medium. (Photograph © P. Lepoint and H. Maraite.)

Plate 12. Asci of *Gibberella xylarioides*, the coffee wilt pathogen, containing two celled ascospores (arrowed). (Photograph © J. Flood.)

Plate 13. Failure to produce orange pigmentation, a characteristic used to rapidly differentiate biological species BS 1 from other representatives of the G. *xylarioides* complex. In this case, *MAT-1 C. canephora*-associated (BS 2, orange colonies) mating type tester strains were confronted with a *MAT-2 C. arabica*-associated *G. xylarioides sensu lato* strain (BS 1, white colony) on carrot agar. Incompatibility between the two BS is denoted by the absence of perithecia or protoperithecia along the confrontation zones. (Photograph © P. Lepoint and H. Maraite.)

Plate 14. Production of perithecia along confrontation zones of *G. xylarioides sensu lato* strains of opposite mating type and of the same biological species (BS). (Photograph © P. Lepoint and H. Maraite.)

15

16

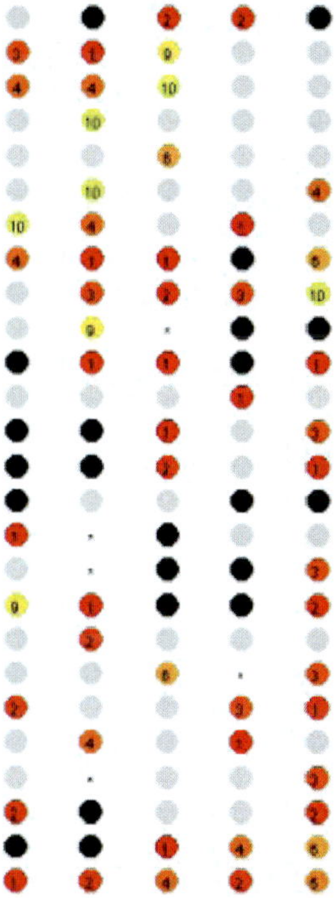

Plate 15. Purple black *G. xylarioides sensu lato* perithecia producing an orange cirrhus of ascospores on carrot agar medium when *C. canephora* associated strains of opposite mating type are crossed. (Photograph © P. Lepoint and H. Maraite.)

Plate 16. Graphical representation of the spatial and temporal development of coffee wilt disease on coffee trees on a farm in Mayuge, Uganda. This depicts the point, during a 17 month study involving ten assessments, at which external disease symptoms were first observed. Points where trees were missing at the outset of monitoring, dead trees and trees that did not develop symptoms throughout the study are denoted by white, black and grey spots respectively. Trees that developed symptoms in the earlier, intermediate and latter stages of the study are denoted by red, orange and yellow spots respectively, the numbers denoting the actual assessment point (1-10) at which symptoms were first observed. (Illustration courtesy of M. Rutherford.)

17

18

19

Plate 17. Host specificity of *Gibberella xylarioides* isolates collected from *Coffea arabica* and *C. canephora* on seedlings of the respective host species. (Photograph courtesy of A. Girma.)
Plate 18. Host specificity of *Gibberella xylarioides* isolates from *Coffea arabica* (G3P22), *C. canephora* (CAB003, TZ009) and *C. excelsa* (DSMZ 62457) on seedlings of *C. arabica* (*cv.* SL28) in growth room. (Photograph courtesy of A. Girma.)
Plate 19. Specificity of *Gibberella xylarioides* isolate DSMZ62457 (an historical Excelsa strain), G3P22 (Arabica strain) and CAB003 (Canephora strain) on Catimor seedlings (1570) in growth room. (Photograph courtesy of A.Girma.)

20

21

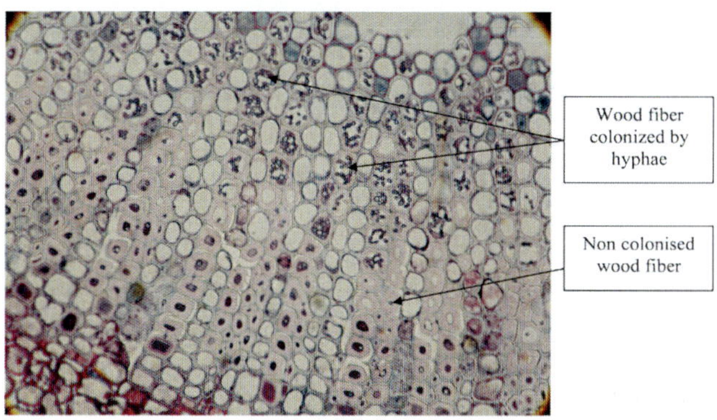

22

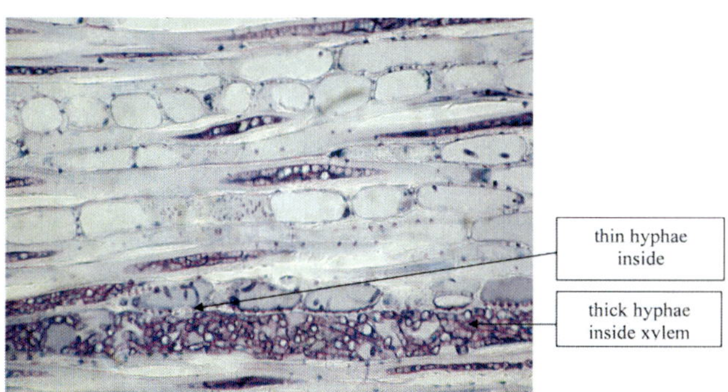

Plate 20. The characteristic unilateral wilting symptoms caused by the historical strain DSMZ 62457 on *C. liberica* seedlings. (Photograph courtesy of A. Girma.)

Plate 21. *Gibberella xylarioides -Coffea canephora* interaction at cellular level (cross section of a 9-month old infected seedling (Photograph courtesy of D. Bieysse.)

Plate 22. *Gibberella xylarioides-Coffea canephora* interactions at cellular levels (arrow hyphal colonization of vascular tissues of 18-month old seedlings 45 days after inoculation. (Photograph courtesy of D. Bieysse.)

23

24

25

Plate 23. Screening *Coffea arabica* collections by seedling inoculation test in the green house in Ethiopia (Jimma). (Photograph courtesy of A.Girma.)

Plate 24. *Coffea arabica* seedlings re-inoculated to verify the resistance in the screen house. (Photograph courtesy of A.Girma.)

Plate 25. *C. canephora* seedlings that survived (resistant) CWD infection after the first bout of inoculation. Left foreground, susceptible plants: right foreground, survivors. (Photograph courtesy of P. Musoli.)

26

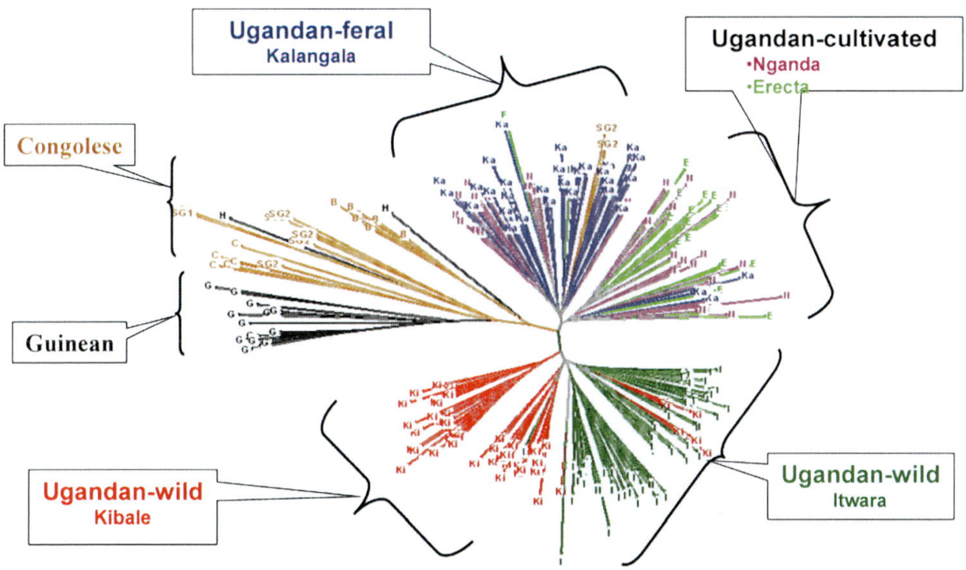

27

Plate 26. Diversity of Ugandan *C. canephora* and other diversity groups. (Illustration courtesy of P. Musoli.)

Plate 27. Eracta and nganda *C. canephora* trees. (Photograph courtesy of P. Musoli.)

28

29

30

31

Plate 28. Uprooting and burning is the most effective eradication strategy for CWD infected trees. (Photograph courtesy of N. Phiri.)

Plate 29. FFS farmers practicing pruning technique. (Photograph courtesy of N. Phiri.)

Plate 30. A coffee farm mulched with dry grass which suppressed weeds. (Photograph courtesy of N. Phiri.)

Plate 31. A coffee farm planted to *Desmodium intotum* which suppressed weeds. (Photograph courtesy of N. Phiri.)

32a

32b

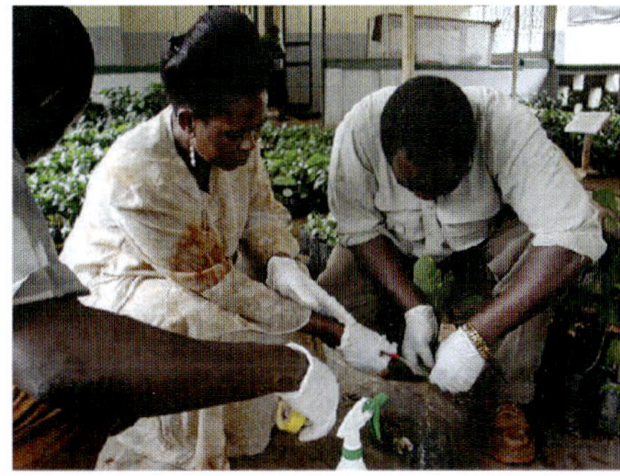

33a

33b

34

35a

35b

35c

Plate 32. (a and b) TaCRI team receiving training at Kituza. (Photograph courtesy of H. Mutenyo.)

Plate 33. (a and b) Spreading CWD technologies through participatory approach.

Plate 34. Farmers carrying out agro-ecosystem observation and analysis in small groups.

Plate 35. Media interviews with (a) Ethiopian researchers, (b) farmers and (c) CABI scientists.

conducted in four regions confirmed to have CWD, namely, the Southern Nations Nationality Peoples, Oromiya, Gambella and Amhara. Households with a CWD severity of at least 10% were randomly selected from each of the regions. A total of 137 households were interviewed in Ethiopia. In DRC, the survey was conducted in North Kivu and eastern provinces where a total of 436 households were randomly selected and interviewed.

Structured questionnaires were used to collect data from all the selected households in all the countries between March 2002 and January 2003. Data collected included household characteristics, farm characteristics, farmers' perceptions of CWD, awareness and ability to identify the disease, severity of CWD and its progression on farms, coping strategies by farmers, income changes, input use, asset portfolio and changes in household expenditure. Data collected were analysed using descriptive statistics, regression analyses and chi-square tests to establish relationships. Analysis was conducted using the Statistical Package for Social Scientists. Information from other minor studies on the disease carried out in individual countries has also been incorporated for comparison purposes and to enrich the chapter.

6.2. Farmer Perceptions of CWD in ECA

Farmer perceptions of CWD varied across the region. Overall, awareness was moderate (40.6%) except for Uganda (90%), where the disease had been particularly severe and the majority of farms (90.3%) had CWD. Its early detection in the country was also partly responsible for this high awareness (Hakiza, 1995, unpublished). In DRC, farmers' knowledge of the disease is moderate, and about 40% of the interviewed farmers know CWD. Awareness of CWD was similar in Ethiopia (17%) and Tanzania (15%). Although CWD is caused by a fungus, some farmers attributed the disease to insects, soil-borne pathogens, nematodes, ants, soil nutrient depletion and the environment. There was lack of consistency on when farmers thought the disease spreads most, whether in the rainy season or in the dry season. In Tanzania and Uganda, more than half of the farmers reported that the disease spreads more during the dry season, whereas more than half of the farmers in Ethiopia indicated the opposite, that the disease spreads more during the wet season. About 34.8% of the farmers in DRC indicated that the disease spreads most during dry weather, whereas 22.2% thought that the disease spreads during wet periods. The remainder of the farmers did not know when the disease spreads most. Dry weather resulting in moisture stress makes coffee plants more vulnerable to CWD, which may lead to more observable effects during the dry season compared to the wet season. The fungus is however known to germinate and spread more during the wet rather than the dry season. All farmers interviewed noted that CWD was causing major damage to the coffee crop on their farms. On the trend of the disease since farmers first noticed it, majority of farmers in all the countries (93.8% in Tanzania, 73.7% in Uganda and 97% in Ethiopia) indicated that the disease was increasing. This meant that control measures would be necessary to reduce crop losses.

6.3. Changes in the Importance of Coffee as a Source of Income

Prior to the re-emergence of CWD, there was already significant variation in the importance of coffee as a source of income across the region. Overall, it was the major source of income for 70% of coffee farmers, although the importance was much lower in Uganda (15%) compared to DRC (95%), Ethiopia (91%) and Tanzania (95%). The relative contribution of coffee to household income in the region has, however, declined by an average of 12% following the onset of CWD, and currently, only an average of 58% of coffee farmers consider coffee to be their major source of income. It is expected that if CWD is not managed, the importance of coffee as a source of income is likely to decline further. Similarly, the percentage of farmers with coffee as their only source of income has fallen from 50% to 44% across the region. There is a general decline in earnings from coffee in the ECA region, which is attributed to CWD. The fall in number of farmers that had coffee as the main source of income and decline in coffee incomes was aggravated by the fall in world coffee prices during the period.

Before the onset of CWD in Tanzania, coffee was the major source of income for 95% of the farmers and the second most important source of income for the remaining 5%. After the onset of CWD, the percentage dropped to 91%, with some of the farmers opting for fishing and fish trade and growing bananas and beer brewing. The alternative to coffee growing adopted by farmers depended on level of accessibility to the enterprise. Whereas the percentage of farmers with only one source of income was 58% before CWD, this fell to 53% after CWD. The small decline in the relative importance of coffee in Tanzania is because most of the coffee grown here is arabica coffee, which was not affected by CWD strain in the country. Only the CWD strain that attacks robusta coffee is present currently in Tanzania. Hence, the arabica coffee productivity masked the effect of the CWD.

In Ethiopia, the percentage of farmers who rated coffee as the most important source of income was 91% before CWD but fell to 84% after CWD. Eighty percent of the farmers indicated that income from coffee was decreasing, whereas 10% indicated that it was increasing. The farmers that reported increases in income may have escaped the effect of CWD and also had better access to marketing services. Another 1.5% reported that there was no change in the income from coffee. Considering the proportion of income from coffee for households, 33% reported that the share when compared to other sources of income was decreasing, 51% reported that it had not changed while 15% reported that it was increasing.

The contribution of coffee to household income greatly diminished following the CWD epidemic in Uganda. Twelve percent of the farmers interviewed rated coffee as the most important source of income. Livestock production and trading in non-agricultural goods, such as household consumables, became more important as income sources following the CWD epidemic, implying that these are important strategies adopted as a means of coping with CWD in Uganda. The reason for increased livestock keeping may be because of its importance as an alternative income source and a store of wealth (asset). Before CWD, coffee plantations served as an asset and income source held by many farm household currently investing in livestock, especially cattle, as an additional asset. This created an opportunity for increasing the integration of crops and livestock production to exploit synergistic interactions between the two enterprises to achieve sustainable agricultural growth. In DRC, coffee was generating about 95% of income for farmers before re-emergence of CWD. However, a combination of fluctuating and falling prices of coffee in the world market and CWD has reduced the contribution of coffee to farmers' income. Coffee now contributes 20% of income of the farmers.

6.4. Change in Coffee Production and Income

Production of coffee at the farm level fell by an average of 35.0% after CWD infestation, which was reflected in significant ($P < 0.01$) reduction of income across the region. Decline in production for individual countries were 38.6%, 37.0% and 29.4% for Uganda, Ethiopia and Tanzania, respectively.

There are a number of factors that may explain yield loss at the farm level. The fall in world market prices of coffee has significantly reduced the amount of money from coffee that farmers take home (Charveriat, 2001). As a result of this, management of coffee has declined due to a reduction in the amount of money that farmers have for purchase of inputs and for carrying out management practices on their coffee. The yield losses given above are therefore a result of a multiplicity of factors and may not be entirely explained by the incidence or severity of CWD. To determine how much of the yield loss was attributable to CWD, a model was developed using data for farmers' perceptions of the changes in yield before and following the onset of CWD and data for the percentage of trees lost to CWD. This showed a direct correlation between the changes in yield and the percentage of trees lost due to CWD as depicted for Tanzania in Fig. 6.1.

The percentage annual yield loss attributable to CWD was then calculated as follows:

$$\frac{\text{Number of trees lost to CWD per annum}}{\text{Initial number of trees per farm}} \times 100$$

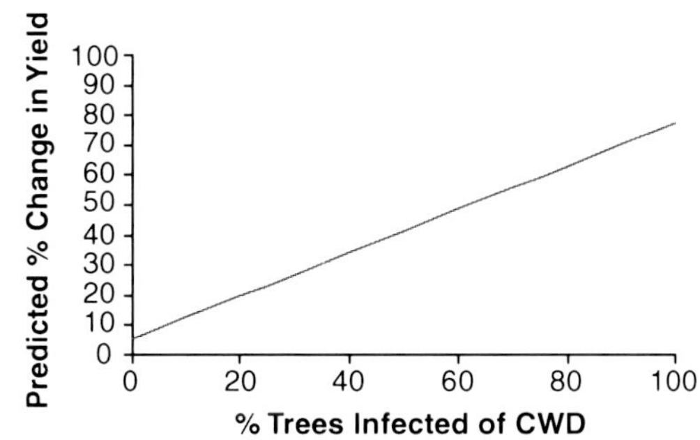

Fig. 6.1. Predicted yield loss based on the incidence of CWD in Tanzania.

Using this method, annual reductions in coffee yield of 7.4%, 1.6% and 2.6% were calculated to be directly attributed to CWD in Uganda, Ethiopia and Tanzania (Kagera Region only), respectively. Assuming that the loss due to CWD is constant and taking into account the average number of years the farmers have had CWD, this equates to a total yield loss of 39%, 16.7% and 10.5% in Uganda, Ethiopia and Tanzania, respectively. It should be noted that coffee trees that had been recently planted and were not yet 'bearing' were excluded from the calculations.

In Tanzania, the amount of coffee produced fell significantly ($P < 0.01$) after CWD infestation from an average of 950 to 686 kg/ha, which is a change of 28.0%. A comparison of the mean coffee production before and after CWD showed significant difference ($P = 0.001$). Coffee income fell from an average of US$531 to US$132, which is a drop of 75%. Comparing the mean income before and after CWD, the difference was found to be significant ($P = 0.001$).

In Uganda, coffee income reduced as a result of CWD. Among the small farmers, income from coffee significantly fell after CWD for all severity levels (low, medium and high) and so did the quantity of coffee produced and sold, except in the low-disease-severity areas where the change in coffee production and sales was insignificant. The area under coffee fell significantly after CWD in the high-severity areas. At medium- and low-disease-severity levels, the area under coffee rose, but the rise was statistically insignificant. Overall coffee production reduced significantly after CWD (Fig. 6.2).

Changes in area, production, sales and income for coffee were separately analysed for small-scale farmers (<5 acres) and large-scale farmers (>5 acres). These changes were analysed at different levels of disease severity (low [L], medium [M] and high [H]), wherein below 5% severity is considered low, 5% to 20% severity is considered medium and above 20% severity is high (Table 6.2). In DRC, coffee production experienced a sharp fall of about 71.5% in

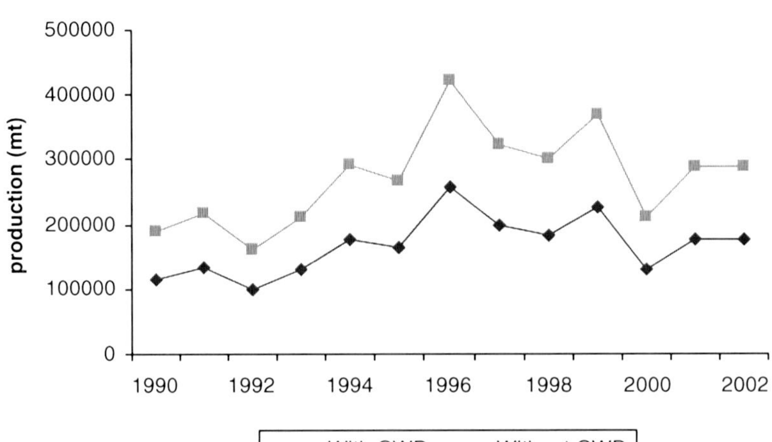

Fig. 6.2. Impact of CWD at national level in Uganda.

Table 6.2. Coffee area, production, sales and income before and after CWD infestation for small farmers (<5 acres of coffee).

Category	Incidence level	Before CWD	After CWD	*T* value
Area (acres)	L	1.42 (0.91) (*n* = 16)	1.73 (1.35) (*n* = 18)	–0.78, NS
	M	1.87 (1.34) (*n* = 36)	1.88 (1.39) (*n* = 39)	–0.03, NS
	H	2.06 (1.35) (*n* = 160)	1.19 (1.13) (*n* = 139)	6.0***
Quantity produced (60-kg bags)	L	7.19 (3.16) (*n* = 14)	6.58 (4.74) (*n* = 16)	0.4, NS
	M	18.13 (17.79) (*n* = 32)	11.78 (11.25) (*n* = 33)	1.72, NS
	H	19.17 (16.98) (*n* = 154)	5.32 (5.25) (*n* = 118)	8.5***
Quantity sold (60-kg bags)	L	7.59 (2.90) (*n* = 13)	7.92 (7.15) (*n* = 17)	–0.16, NS
	M	16.71 (15.51) (*n* = 29)	10.27 (8.02) (*n* = 32)	2.06**
	H	19.12 (16.95) (*n* = 143)	5.05 (5.06) (*n* = 112)	8.49***
Income (Ushs)	L	338,542 (278,137) (*n* = 12)	101,800 (97,626) (*n* = 15)	3.08***
	M	475,136 (317,844) (*n* = 25)	163,559 (162,458) (*n* = 29)	4.63***
	H	714,202 (710,740) (*n* = 123)	30,199 (27,508) (*n* = 90)	9.12***

Figures in parentheses are standard deviation.
*, ** and *** imply statistically significant differences between means at 10%, 5% and 1% levels of significance, respectively; NS = implies not significant; *n* = number of observations.

North Kivu province and 78% in eastern provinces. This is attributed to an average destruction of 487 and 1029 coffee trees per farm in North Kivu and eastern provinces, respectively.

A Uganda Coffee Development Authority (UCDA) survey confirmed that all robusta-producing districts have been affected by CWD. In some cases, entire fields of coffee have been destroyed. A 2003 review study estimated that of the total of 240,000 ha in all 21 robusta-growing districts, 122,400 ha have been infected. UCDA estimated a loss equivalent to 61,200 t of coffee, which is around 40% of the output in recent years (Baffes, 2006). The number of living trees in Uganda is smaller than it was in the early 1990s (EPL, 2005). However, in its effort to contain the disease, Uganda has undertaken a replanting programme where farmers who uproot infected trees are given coffee-planting materials. Likewise, coffee has been introduced in northern Uganda as a cash crop (UCDA, 2004).

In the case of Ethiopia, there was a 37% loss in yield from coffee, which fell from a yield of 481 to 303 kg/ha after CWD. The income from coffee fell by 67% from US$553 to US$181 after CWD. The mean yield before and after CWD were significantly different ($P < 0.05$). These yield differences also translate into production (Fig. 6.3).

Geiser *et al.* (2005) also asserted that there is a substantial loss in coffee yield in Ethiopia, which is to a large extent determined by the wide spread of CWD. The income before and after CWD were also found to be significantly different. The yield loss attributable to CWD was then calculated by multiplying the average trees lost due to CWD and the average yield per tree. This gave a yield loss of 23.1%. Thus, of the 37% change in yield, 23.1% was a result of CWD.

In DRC, CWD infection caused a decrease in funds available for health and education resulting in high reduction of schooling population and increase in infant death rate. Similarly, malnutrition and the number of homes living below poverty levels increased. To compensate for fall of revenues formerly generated by coffee, the farmers in DRC increased land for food crop production for sale at the expense of household food consumption. CWD destroyed almost 28,853 ha formerly setaside for robusta in DRC. This was accompanied by the abandoning of much of the infrastructure leading to immobilization of huge amounts of capital meant to contribute to socio-economic development through employment. In 1989, territories of Beni and Lubero in North Kivu had 24 coffee-processing factories with a capacity of 452 t per day and 24 warehouses with total capacity to stock 50,900 t. Currently, more than 90% of employment that these infrastructures were creating is non-existent.

Using the annual percentage yield loss (computed in section 5.4) for each country, the impact of CWD at the national level was calculated using FAO coffee production figures for 2002 (FAO, 2004) and International Coffee Organization indicator prices for 2002 (ICO, 2004). Annual losses attributed to CWD were 14,573 t amounting to US$9,644,279 in Uganda and 3360 t

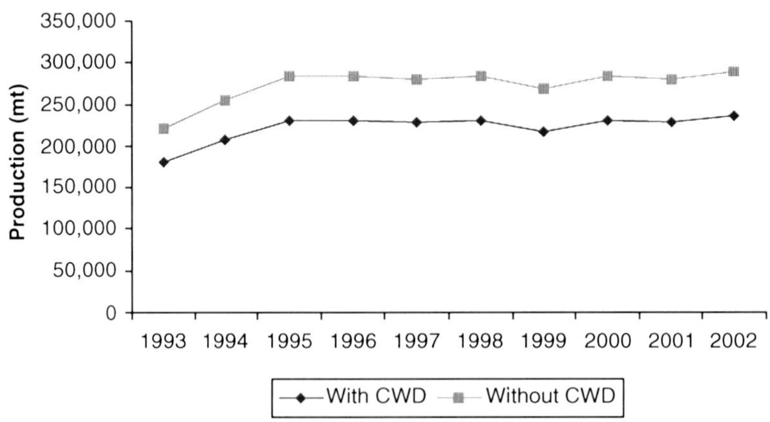

Fig. 6.3. Impact of CWD at national level in Ethiopia.

amounting to US$3,750,976 in Ethiopia. In the case of Tanzania, the national level impact of CWD was calculated using the production figures for the Kagera region because this was the only region where CWD was found. The annual yield loss for Tanzania was 299 t amounting to US$197,551. CWD has caused a disaster to the coffee industry at national level in DRC leading to a vicious circle of poverty. There has been a reduction in investment and resource mobilization leading to stagnation in the level of production and productivity and a fall in revenues. The annual loss in DRC is 68,000 t of coffee, which translates to US$46.6 million.

6.5. Change in Input Use and Management of Coffee

Across the region, there has been a significant decrease ($P < 0.01$) in the use of inputs on coffee. Prior to the onset of CWD, an average of 24.2% of farmers were using inputs on coffee (19.4% in Uganda and 29% in Tanzania). Of the farmers that were using inputs, 46.2% were using organic manure, 15.9% were using pesticides, 43.5% were using herbicides (only in Uganda) and 13.7% were using fertilizers. After the onset of CWD, the percentage of farmers using inputs on coffee across the region decreased to only 7.1% (11.2% in Uganda and 3% Tanzania). Of those farmers still using inputs, use of manure fell from 46.2% to 14.2% and the use of fertilizers also decreased from 13.7% to 5.8%. Table 6.3 provides changes for individual countries.

Labour inputs into coffee production fell by an average of 33.5%. This is because as income from coffee declines, farmers are discouraged from investing in the crop. A few farmers uprooted diseased plants (except in Ethiopia) to remove coffee and replace it with other crops such as pineapple, bananas and vanilla. In Uganda, the number of households spending on non-labour inputs for coffee production after CWD reduced by 52% in the high-disease-severity areas and by 39% in the medium-disease-severity areas. Most farmers in Ethiopia did not use any inputs on coffee either before or after CWD. There was no use of fertilizer, pesticides or insecticides. The only input on coffee was labour. Before CWD, family labour input on coffee was an average of 1625 h per household per year, but after CWD, this decreased to 1297 h per household, which is a 20% decrease. There was however a drop of 69% in

Table 6.3. Changes in level of input usage (%) in different countries.

	Ethiopia		Tanzania		Uganda	
Type of input	Before CWD	After CWD	Before CWD	After CWD	Before CWD	After CWD
Organic manure	0.0	0.0	62.0	2.0	30.4	0.0
Pesticides	0.0	0.0	31.0	31.0	5.8	2.0
Herbicides	0.0	0.0	0.0	0.0	43.5	45.0
Fertilizers	0.0	0.0	7.0	7.0	20.3	0.0
Family labour	100.0	80.0	100.0	83.0	100.0	75.0

hired labour from 4031 to 1244 h per household per year. Overall, there was a 51% drop in the labour allocated to coffee.

Management practices on coffee in Tanzania include weeding, mulching, application of manure, stumping, pruning and spraying. Before CWD, 93% of the farmers weeded their coffee. This percentage did not change even after the onset of CWD. However, the percentage of farmers who practiced mulching fell from 47% to 38%. The highest reduction was in the percentage of farmers who were pruning, which fell from 93% to 33%.

Management practices on coffee fields in Uganda include weeding, mulching, pruning/de-suckering and spraying. Stumping is usually done once in 5 to 6 years. Although stumping is one of the recommended practices, farmers are reluctant to use it for fear of loosing the crop for several seasons. Only a small proportion of farmers mulch their coffee. Coffee management practices in the affected countries are declining due to neglect of coffee.

6.6. Asset Liquidation as a Result of CWD

Twenty-seven per cent of the coffee farmers throughout the region liquidated some of their assets to meet their financial obligations as a result of reduced income from coffee. Assets included land (16% of farmers), livestock (12% of farmers), electronic equipment (10% of farmers) and bicycles (12% of farmers). Despite the sale of assets, some farmers still failed to meet their financial obligations, the most common of which was completion of houses that they had started to build prior to the re-emergence of CWD and school fees for their children. The money obtained from the liquidated assets was used mainly on investment in non-crop-farming enterprises such as starting poultry farms, transportation business, house construction and for solving household needs such as purchasing food and paying for medical expenses and financing burial arrangements.

After the onset of CWD, 39.4% of farmers in Tanzania disposed some of their assets. Of these, 5% disposed more than one asset. Most of the farmers sold livestock, electronics and bicycles/motorcycles. The value of assets sold varied from US$4 to US$1250. The liquidation of the assets is mainly due to reduced income from coffee and increased expenditure on CWD.

In Uganda, a few farmers, particularly in the high-disease-severity areas, liquidated some of their assets following the outbreak of CWD so as to cope with the disease. The most commonly liquidated assets were land, followed by communication equipment then bicycles and livestock of all types. The money obtained from the liquidated assets was mainly used on investments in non-crop-farming enterprises such as starting poultry farms, purchasing motor bike to start transportation business, constructing houses and solving household and domestic needs such as purchasing food and paying for medical expenses, school fees and financing burial arrangements. The value of assets sold varied from US$20 to US$857 in Uganda. In addition, 16.6% of coffee farmers had started borrowing money after CWD to meet their short-term financial requirements, whereas 7% rented out land.

Twenty-six per cent of households interviewed in Ethiopia liquidated their assets as a result of CWD. Most of the assets liquidated were livestock including cows, oxen, sheep, goats, donkeys and chickens.

6.7. Changes in Household Expenditure Using Income From Coffee

Information was sought on changes in proportion of expenditure for various items that were coming from coffee before and after CWD to measure the impact of CWD on farmers' family welfare. Patterns of household expenditure with income from coffee changed following the general reduction in income from coffee, although the trends were not consistent across the region. In Uganda, there was a significant reduction in the proportion of expenditure on all items (education, health, food consumption, etc.) except labour following the outbreak of CWD. In Ethiopia, the proportion of income from coffee spent on household items decreased significantly ($P < 0.01$) from 85.6% to 81.2% following the onset of CWD. Income from coffee was used primarily on food, clothing and health, in order of priority, with some farmers depending on it for almost all of their food supply. The proportional reduction in expenditure on food is leading to an increase in food insecurity, especially protein. This in turn negatively impacts on productivity. In Tanzania, there was very little change in the amount spent on items such as food and health, but the proportion spent on education increased from 33.1% to 41%, and there was a decrease in the proportion spent on labour (from 13.9% to 8.5%), investments (from 26.3% to 9.1%) and leisure (from 13% to 6%). Thus, following the onset of CWD and the subsequent reduction in income from coffee, households in Tanzania opted to cut down on non-essential items and reallocate the available income to essential items.

6.8. Farmer Coping Strategies and Factors Influencing Them

Following the onset of CWD, farmers adopted a range of different strategies for coping with the effects of the disease. These included the following:

- Re-allocating labour to other on-farm activities
- Diversifying into other crops not previously grown including bananas, beans, cassava, fruit, groundnut, maize, millet, sorghum, onion, peas, sweet potato, tomato, vanilla and yams
- Starting other non-farm activities including brewing, fishing and trade
- Opening up new coffee fields
- Replacing infected coffee trees
- Renting out coffee fields
- Borrowing money from accessible sources.

There were notable variations in the relative importance of different coping strategies across the region. Farmers in Uganda coped with CWD by re-allocating labour to other on-farm activities (57.6%), re-allocating labour to off-farm activities (22.4%), abandoning coffee fields after CWD (18.2%) and renting out coffee fields (1.8%). In Ethiopia, however, farmers gave their coping strategies as replanting coffee (76.5%) and replacing infected coffee trees (23.5%), which is distinguished from replanting in that replacing means planting new coffee at the same place where the uprooted coffee was. This is common in Ethiopia as farmers often attempt to replant seedlings from forest or other coffee fields. Unlike other countries in the region, farmers in Ethiopia did not start other off-farm activities nor diversify into alternative crops. This is because of the socio-cultural significance of coffee in Ethiopia, which is deeply rooted in traditional culture and sentiment and farmers do not easily resort to production of other crops. Regardless of what happens to productivity, farmers keep their coffee at least for their own consumption. Moreover, production of other crops is highly constrained by wildlife because most of the coffee areas have forests harbouring wild animals. In Tanzania, the main coping strategies included diversifying into other crops not previously grown (33%); starting off-farm activities such as brewing, fishing and trading (19%); opening up new coffee fields (4%) and replacing infected coffee trees (48%). In addition to the yield loss and decline in income from coffee, farmers have been obliged to spend money especially for hiring labour to uproot and burn infected trees. Some farmers also incurred additional costs to purchase chemicals and other control methods they felt would control the disease. In Uganda for example, 58% of farmers adopted uprooting and burning of infested trees as a control measure. Other technological control measures included the use of ash, urine and tobacco.

Socio-economic factors influencing the choice of coping strategies adopted by coffee farmers were evaluated using a binary logistic regression model for each country. In Uganda, the choice of re-allocating labour from coffee to other on-farm activities as a coping strategy for CWD was influenced by the number of years of experience of coffee farming, the yield of coffee prior to the onset of CWD and whether coffee was the only source of income for the household. Switching to non-farm activities as a coping strategy for CWD was influenced positively by the income of coffee prior to the onset of CWD and the proportion of household labour inputs into coffee. Relatively higher coffee income before CWD meant the farmer was more likely to switch to non-farm activities after the onset of CWD. Higher household labour input into coffee prior to CWD meant that farmers were more likely to switch to non-farm activities after infestation of CWD. The likelihood of adopting non-farm activities as a coping strategy was also increased by the proportion of trees lost and quantity of coffee produced before CWD.

In Ethiopia, the choice of coping strategy was influenced by the level of education of the farmer. Farmers with a lower level of education were more likely to replant coffee (97%) than farmers with higher levels of education (P = 0.012). Although replacing coffee trees is not common in Ethiopia, farmers with no education were more likely to replace (29%) than farmers who had primary school or higher levels of education (P <0.05).

Coping strategies were also influenced by two biological factors, namely, severity of CWD on-farm and the yield of coffee prior to the onset of CWD.

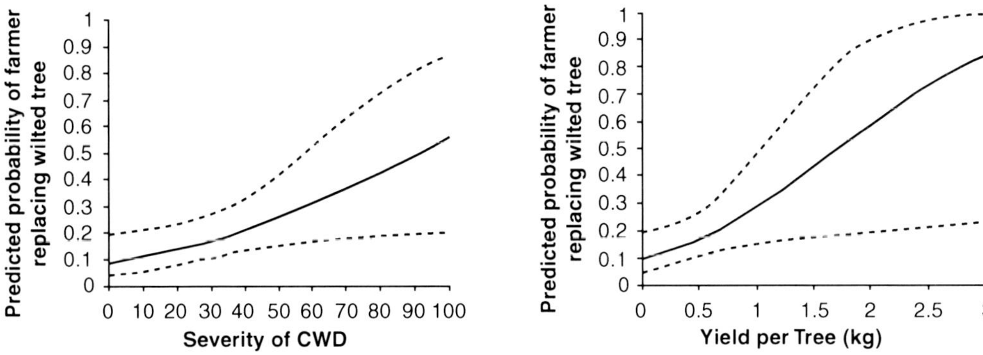

Fig. 6.4. Effect of severity of CWD (left) and coffee yield (right) on the predicted probability of a farmer replacing a wilted coffee tree in Ethiopia.

The higher the severity of CWD, the higher the probability of the farmer replacing the infected coffee trees ($P = 0.024$) (Fig. 6.4). Similarly, the higher the yield before CWD, the higher the probability that farmers would replant their coffee ($P = 0.037$; Fig. 6.4).

In Tanzania, the likelihood of a farmer choosing non-farm activities as a coping strategy was significantly reduced as the age of the farmer increased (Fig. 6.5). Of the non-farm activities, fishing was the most practiced, and because it was done mainly at night, it was particularly suitable for young and/or unmarried men as compared with the older ones. Farmers who had been growing coffee for a longer period of time were also more likely to grow alternative crops (Fig. 6.5). Rather unexpectedly, the length of time CWD had been present on a farm was negatively correlated with the likelihood of a farmer switching to alternative crops (Fig. 6.5).

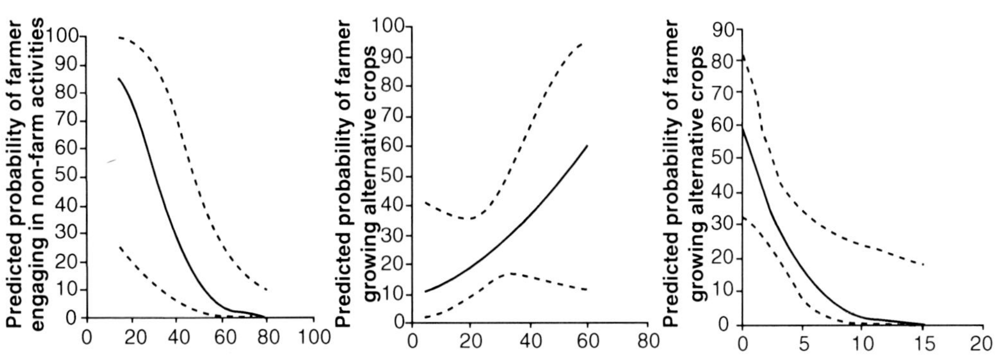

Fig. 6.5. Effect of age of farmer (left), number of years of coffee farming (centre) and number of years of CWD on farm (right) on predicted probability of farmer diversifying from coffee to alternative crops as a result of CWD in Tanzania.

6.9. Conclusions

Coffee is an important commodity in the region because of its contribution to income, foreign currency, employment and livelihood. Coffee production and productivity are affected by CWD, which is a serious coffee disease in the region, although its effect is not uniform and some places are affected worse than others. CWD appears to have a significant impact, but this is confounded by impacts on the coffee business at the global level. Farmer understanding of the disease and its control methods are still low in the region, although higher in Uganda where the disease has been for long and widespread. Low understanding of the disease coupled with financial limitations and knowledge of CWD occurrence has resulted in limited farmer capacity to control CWD in the region. Low technical know-how and the low availability of a diverse range of affordable control methods have also contributed to a situation where there is little attempted management of CWD, and this has led to reduced coffee yields and overall coffee production. The importance of coffee as a major source of income in the region declined due to the effects of CWD. Use of inputs for coffee fell significantly in all the countries as did use of labour for coffee. Similarly, production of coffee at farm level fell significantly in all the countries. Following the re-emergence of CWD, income from coffee declined by over 50% in the affected countries. This has led to losses at farm, household and national levels. The livelihood of farmers has therefore declined further because they were forced to sell their assets and restrict expenditure to basic requirements only. Reduction in regular income as a result of reduced income from coffee led farmers to liquidate their assets after CWD to meet their financial obligations. Household expenditure fell in all the affected countries after CWD. Many coping strategies were adopted for CWD. There were however notable variations in their relative importance across the region.

The rating of coffee as an important source of income in the region and individual countries means that efforts should be directed at managing CWD and increasing coffee production. CWD control requires concerted efforts. An integrated approach involving farmers and the institutions concerned with crop protection in the respective countries is essential. Farmer awareness of the disease has to be increased coupled with an increase in alternative methods of control. Continuously increasing awareness about CWD and how to manage it is likely to lessen the rate of wilt spread. Raising awareness is likely to be important where the severity of the disease is high and for farmers that are most likely to persevere with coffee rather than diversify to other activities. Awareness campaigns through national extension programmes, farmer groups and farmers' field schools are needed. Awareness creation should be associated with networking on how to manage the disease. Similarly, farmer capacities need to be improved by facilitating access to the requisite resources for control of the disease. An appropriate farmers' field school curriculum to empower coffee farmers using participatory approaches to raise awareness of CWD and its management would help reduce CWD. Farmer training and implementation of good husbandry procedures and practices can increase potential for stabilizing the social situation in the long term. In this way, rural incomes can also be improved through value addition.

Host resistance would offer the best prospect for control and demand very little effort from the farmers. This means that a long-term solution to CWD would be development of wilt-disease-resistant varieties. Intensification and acceleration of research for wilt-resistant varieties should receive attention (see Chapter 10). In the short term, breeding should be accompanied with provision of free and clean planting materials. Revival of research to generate a diverse range of control strategies should also be considered.

Acknowledgements

Collection of socio-economic data and information was facilitated by many persons in different organizations in target countries. The country level contacts for the CWD project are highly acknowledged. The coordinator of the coffee wilt project is highly acknowledged for providing assistance in data collection and provision of most of the literature reviewed while preparing this chapter. We wish to thank CABI Africa and other organizations to which the co-authors are affiliated. This work was funded by the European Union and the Common Fund for Commodities. The financial support is highly appreciated.

References

Baffes, J. (2006) Restructuring Uganda's coffee industry: why going back to basics matters. *Development Policy Review* 24, 413–436.

Baffes, J. (2003) Tanzania's coffee sector: constraints and challenges in a global environment. Africa Region working Paper No. 56, June 2003. www.worldbank.org/afr/wps/wp56/htm.

Banque Centrale du Congo (BCC) (2005) 1995–2007 Rapports Annuels d'Activités.

Beintema, N.M. and Solomon, M. (2003) Agricultural science and technology indicators for Ethiopia. ASTI country Brief No. 9. October 2003.

Charveriat, C. (2001) *Bitter coffee: How the poor are paying for the slump in coffee prices*. Oxfam, Oxford.

EPL (The Evaluation Partnership Limited) (2005) *Impact evaluation of the four CFC funded projects in Uganda*. The Evaluation Partnership Limited, Middlesex, UK.

FAO (2004) Agriculture and food trade statistics. www.faostat.fao.org.

Geiser, M.D., Lewis, I.L.M., Hakiza, G., Juba, H.J. and Miller, A.S. (2005) *Gibberella xylariodes* (anamorph: *Fusarium xylariodes*), a causative agent of coffee wilt disease in Africa, is a previously unrecognized member of the *G. fujikuroi* species complex. *Mycologia* 97, 191–201.

GoU (Government of Uganda) (2001) The path forward in Uganda's coffee sector. A report prepared for the presidential conference on export competitiveness. Prepared by the complete project and Alice Agowa (Uganda coffee) on behalf of the European commission, February 2001.

ICO (2004) Trade statistics. Available at: www.ico.org/frameset/traset.htm.

Karanja, A.M. and Nyoro, J.K. (2002) Coffee prices and regulation and their impact on livelihoods on rural community in Kenya. Tegemeo Institute of Agricultural Policy and Development. Working paper, 64 pp.

Oduor, G., Simons, S., Phiri, N., Njuki, J., Poole, J., Pinard, F., Kyetere, D., Hakiza, G., Musoli, P., Lukwago, G., Abebe, M., Tesfaye, A., Kilambo, D., Asiimwe, T. and Munyankere, P. (2003) Surveys to assess the extent and impact of coffee wilt disease in East and Central Africa. Final Technical Report EU contract No. ASA-RSP/CV-006. CAB International, Egham, UK.

Oxfam (2002) Crisis in the birth place of coffee. Oxfam International research paper, September 2002. Available at: http://www.maketradefair.com/en/assets/english/coffeecrisisKafaEthiopia.pdf.

UCDA (2001) UCDA Annual Report, Industrial Graphics Limited, Entebbe, Uganda.

UCDA (2004) UCDA Annual report, Industrial Graphics Limited, Entebbe, Uganda.

7 Biology, Taxonomy and Epidemiology of the Coffee Wilt Pathogen *Gibberella xylarioides* sensu lato

M.A. Rutherford,[1] D. Bieysse,[2] P. Lepoint[3] and H.M.M. Maraite[3]

[1]*CAB International (CABI), Europe-UK, Bakeham Lane, Egham, Surrey, TW20 9TY, UK*

[2]*Centre de Coopération Internationale en Recherche Agronomique pour le Développement (CIRAD), TA41/K, Campus International Baillarguet, 34398 Montpellier, Cedex 5, France*

[3]*Unité de Phytopathologie, Université Catholique de Louvain (UCL), Croix du Sud 2/3, B-1348 Louvain-la-Neuve, Belgium*

7.1. Introduction

Coffee wilt disease (CWD) was first reported on 'excelsa' coffee (*Coffea liberica* var. *dewevrei*, formerly *Coffea excelsa*) in Central African Republic (CAR) in 1927 (Figueres, 1940). Although subsequent epidemics in CAR, Côte d'Ivoire and Democratic Republic of Congo (DRC) between the 1920s and 1960s caused extensive damage to *C. liberica* and *Coffea canephora* (robusta coffee), the disease was successfully reduced mainly by sanitation and elimination of the most susceptible genotypes to the level of a minor disease (Fraselle, 1950; Jacques-Felix, 1950; Saccas, 1951; Jacques-Felix, 1954; Stewart, 1957; Meiffren, 1961; Kranz, 1962) (see also Chapter 9). However, renewed and devastating outbreaks on *C. canephora* in DRC, Uganda and Tanzania, coupled with worsening problems on *Coffea arabica* in Ethiopia (Kranz and Mogk, 1973; Girma *et al.*, 2001; Girma, 2004; Girma *et al.*, 2006), are major cause for concern for future sustainability of coffee production in Eastern and Central Africa.

Until recently, little was known about the fungus responsible for CWD, *Gibberella xylarioides* Heim & Saccas, in comparison with other fusarial pathogens possessing similar capacity for destruction and economic loss. In 2000, the Regional Coffee Wilt Programme (RCWP) was initiated to manage CWD through the development and implementation of an effective and sustainable disease management programme based principally on host resistance. As part of this programme, extensive multidisciplinary research was undertaken to improve our understanding of the occurrence of CWD and of the biological nature of *G. xylarioides*.

This chapter provides a review of current knowledge of the pathogen, incorporating comprehensive new information acquired through international research efforts supported by the INCO-Dev Programme of the European Community and the UK Department for International Development, Crop Protection Programme.

7.2. The Cause of CWD

CWD has been attributed to the fungus *G. xylarioides* Heim & Saccas (anamorph *Fusarium xylarioides* Steyaert; syn. *Fusarium oxysporum* forma *xylarioides* [Steyaert] Delassus). Steyaert (1948) described the anamorph, *F. xylarioides*, as a new species causing wilt of coffee (*C. excelsa*).

7.3. Symptoms

The external symptoms exhibited by coffee plants affected by CWD are generally similar to those of other vascular wilt pathogens. Symptoms may appear at any stage of crop growth, and the rate at which they develop may vary. Once affected, death of the plant is inevitable, and in mature trees, it usually occurs between 3 and 15 months after first appearance of symptoms. Young plants, however, may be killed within a matter of a few weeks of infection.

In young plants, leaves initially exhibit chlorosis (yellowing), begin to droop or wilt and develop brown necrotic lesions, initially along the margins and veins. Lower leaves tend to be affected first, lesions gradually enlarging until the leaves dry, shrivel and abscise, often within a matter of days. In more mature plants, a general chlorosis, wilt and folding or rolling of leaves are among the earliest symptoms observed. Symptoms are often initially restricted to, or more pronounced on, one part of the tree, a likely consequence of initial infection occurring on one of several main stems. Dark brown or black 'streaking' of shoots, clearly visible on younger green shoots, and discolouration of leaf veins (Plate 5) may also occur. Symptoms again culminate in drying and loss of leaves, constituting a progressive and often unilateral dieback and defoliation that gradually extend to the entire plant. Chlorosis, leaf droop and leaf abscission in particular may be more pronounced under natural conditions of water stress (e.g. dry season, drought), less so with the onset of the rains. Coffee berries that would normally be green may redden as if ripening prematurely but often remain intact on shoots following shedding of leaves.

Other external symptoms include swelling of the trunk and the appearance of vertical or spiralling cracks in the bark of mature trees. Small blackish-brown perithecia of the fungus, similar in appearance to dark soil particles, may be produced in the cracks of the bark (Plate 9), particularly during the rainy season or otherwise moist conditions. As described below, perithecia, produced primarily as sexual reproductive structures, may also constitute a means of survival for the pathogen and are commonly observed on dead and decaying material of plants affected by CWD.

It is important to note that many of the external symptoms associated with CWD are not unique to the disease and may also be caused by other pathogens, including other *Fusarium* species (see 7.10), root rot, physiological disorders, such as overbearing dieback, and adverse environmental conditions, including water stress. However, confirmation of CWD can usually be confirmed by the development of bluish-black staining of the wood directly beneath the bark. Staining is usually most pronounced in the collar region but may extend below soil level and/or towards the apex of the tree. Such staining may be confirmed by careful removal of the bark, by hand or with an implement.

7.4. Disease Cycle and Epidemiology

Although recent research has provided further information on the life cycle of *G. xylarioides* and on disease epidemiology, our understanding of the pathogen remains limited in comparison with many other fusaria and vascular wilt pathogens. Some studies of host–pathogen interactions have been undertaken (Chapter 8), but further in-depth research is required to clarify the primary infection points, investigate mechanisms of tissue colonization following infection, determine what physiological and biochemical responses are elicited by the plant and investigate the existence, role and interactions between genes conferring resistance in the host and virulence in the pathogen.

7.4.1. Host range

G. xylarioides would appear to have a narrow host range, coffee being its primary host and perhaps the only plant on which it is pathogenic. It is clear from observations of the occurrence of CWD in the field (including those of comprehensive and systematic surveys completed across East and Central Africa), the results of pathogenicity testing and isolations made from host substrates that the fungus is prevalent on the two most important commercially cultivated coffee species, *C. arabica* and *C. canephora* (Kilambo *et al.*, 1997; Girma, 2004; Kilambo *et al.*, 2004; Oduor *et al.*, 2004; Tshilenge-Djim *et al.*, 2004; Kilambo *et al.*, 2006) (see also Chapters 3, 4 and 5). The disease was also recently observed on excelsa coffee in Uganda, but only on plants held at a coffee germplasm conservation site (to the authors' knowledge, commercial production of excelsa coffee in Africa is very limited). *G. xylarioides* was reported on wild *C. canephora* exhibiting

CWD symptoms in forest areas of Kibale and Itwara in west Uganda and confirmed as pathogenic to *C. canephora* in subsequent pot-based tests (Bieysse, 2005; Bieysse, 2006).

A range of other crops and weed species commonly found or cultivated in coffee farms affected by the disease in Uganda have been investigated as possible alternate hosts for *G. xylarioides* (Serani, 2000; Kangire *et al.*, 2002; Serani *et al.*, 2007). However, the fungus was only recovered from the roots of the banana cultivar kayinja (syn. 'pisang awak') (Serani, 2000; Serani *et al.*, 2007), often intercropped by farmers alongside coffee for production of banana juice and preparation of beer. This strain was recovered from within the root tissues and found to be pathogenic to *C. canephora* in subsequent host inoculation studies (Serani, 2000; Serani *et al.*, 2007; G. Hakiza, personal communication, 2004). Pathogenicity to banana was not assessed. The extent of similarity, including genetic, between this strain and those recovered from coffee affected by CWD is otherwise unclear.

In DRC, a fungus identified as *G. xylarioides* on the basis of morphological and genetic characteristics (i.e. comparison of *tef* 1α sequences against those held for *G. xylarioides* on the FUSARIUM-SEQFUSARIUM-ID database; Geiser *et al.*, 2004), has been found on cinchona affected by *Phytophthora* canker (A. Buddie, P. Cannon and P. Kelly, personal communication, 2007). As yet, it is unknown whether this fungus is pathogenic to cinchona or indeed to coffee. *G. xylarioides* has also been found on cotton seed (Pizzinatto and Menten, 1991) and on rotting tomatoes obtained from fruit markets in Nigeria (Onesirosan and Fatunla, 1976). However, information supporting these reports is limited, and it is unknown whether these fungi differ morphologically, genetically or otherwise to those currently observed on coffee. Studies to investigate and identify other potential hosts for *G. xylarioides* and to assess the pathogenic nature of the fungus on such hosts are otherwise limited. Insects have been suggested as possible vectors for CWD, but attempts to recover *G. xylarioides* from common pests of coffee (including coffee berry borer, *Hypothenemus hampei*), as well as from bees, termites and other insects found on affected coffee farms, have been unsuccessful (Rutherford and Flood, 2005; G. Hakiza, personal communication, 2004).

7.4.2. Infection and colonization of coffee by *G. xylarioides*

G. xylarioides is considered to be an endemic, soil-inhabiting fungus. Although the life cycle and epidemiology of the fungus are not fully understood, the mechanisms of host infection, colonization and symptom development are thought to be similar to those of other vascular wilt pathogens. Initial entry is considered primarily via the roots and lower stem, facilitated by the presence of wound sites as entry points. These may occur naturally as a consequence of farm management practices (see 6.4.3) or by livestock foraging at the base of the coffee tree. Once within the host tissues, the fungus moves within the vascular system and surrounding tissues, leading to restriction of water transfer through direct physical occlusion with fungal material or by the host plant as a response to invasion. As with other vascular wilts, staining of the wood and roots is probably due to changes in phenol metabolism and may constitute a resistance response by the plant (Pegg, 1981; Beckman, 1987; Hillocks, 1992).

7.4.3. Fungal transmission and dynamics of disease spread

Vegetative and reproductive structures of the fungus, namely, microconidia, macroconidia, ascospores and mycelial fragments, may be readily dispersed by air, water (including rain splash and run-off) and through human activity, perhaps attached to soil and plant debris (Van der Graaff and Pieters, 1978). A profuse production of conidia is noticed on faint mycelium patches developing mostly at the stem base under high-humidity conditions (Fraselle, 1950). Perithecia, often observed on coffee wood under natural conditions and often abundant during the rainy season, may constitute as a means of survival given the sparsity or absence of the thick-walled chlamydospores produced by other fusaria (Flood and Brayford, 1997). Perithecial production has been reported to commence on CWD-affected trees within 2 weeks of the onset of the rainy season and to be more pronounced towards the stem base (Musoli and Hakiza, 2007). Day-to-day agricultural practice, exchange of planting material between growers, dissemination of germplasm by nurseries and distribution centres and transportation of coffee by traders are likely to disseminate fungal propagules and contaminated soil, providing an efficient means of disease spread on a local, national or regional scale (Chapters 3, 4 and 5). Wood cuttings, obtained as a means of vegetative propagation, are also exchanged and sold by growers and nurseries to establish new plantings. Coffee berries are obviously widely distributed and marketed and, if contaminated by *G. xylarioides*, may again provide an effective means of dispersal of the fungus. Isolation of *G. xylarioides* from within coffee berries has been reported in Uganda (Serani *et al.*, 2007).

The machete is an essential implement for coffee farmers in Africa and is used on a daily basis to perform a range of tasks including pruning and weeding. On-station trials confirm that wounding of coffee trees with a machete previously used on infested coffee wood is sufficient to transmit the pathogen to, and result in CWD development on, healthy and mature trees under field conditions. Coffee wood, an important source of fuel for farmers, also constitutes a source of infection for young coffee plants, while infested field soil may remain infective to young plants for several months following uprooting and removal of affected trees (Rutherford and Flood, 2005; Musoli and Hakiza, 2007; Chapter 3, this volume).

Studies to monitor the spatial and temporal development of CWD symptoms at on-station and on-farm locations affected by CWD provide an insight on disease spread under natural conditions (Rutherford and Flood, 2005; Musoli, 2007; Musoli *et al.*, 2008). In one instance, incidence of CWD in mature plants of a range of susceptible clones increased from below 3% (9 plants) to 45% (166 plants) over a 2-year period with, on average, six plants developing external symptoms per month (Rutherford and Flood, 2005). Thirty months later, more than 90% of plants had either been killed or showed symptoms of the disease. The rate of increase in disease incidence also varied from clone to clone, suggesting different levels of susceptibility. The studies indicate that although initial foci of disease (in some cases individual trees) may be randomly distributed, these enlarge over time but in no particular direction (Plate 16). They also suggest that infection of adjacent trees, perhaps through root-to-root contact or short-distance dispersal of fungal material, is the primary means of on-farm disease development (Rutherford and Flood, 2005; Musoli and Hakiza, 2007). Although the point at which individual plants became infected in these studies is unknown and the precise mode of infection is unclear, they provide the most accurate information to date on localized spread of CWD and do confirm the rapid and destructive effects of the disease once introduced. They also have important implications in terms of cultural management and the potential for eradicating CWD through early removal and destruction of plants as soon as symptoms become apparent (see Chapter 9). An illustration of the putative life cycle of *G. xylarioides* is provided in Fig. 7.1.

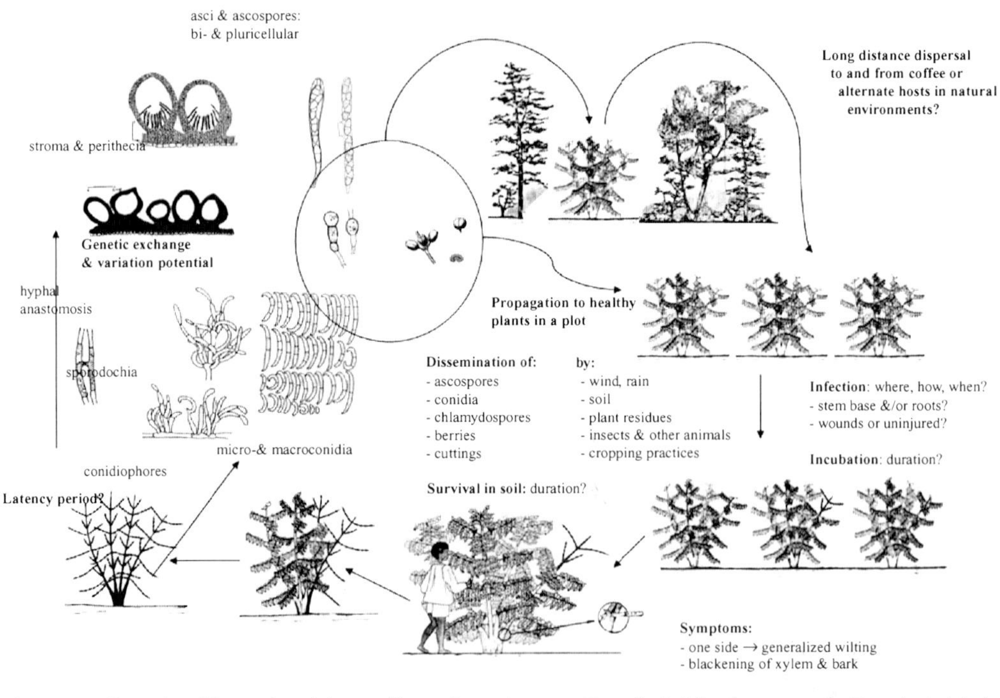

Fig. 7.1. Putative life cycle of the coffee wilt pathogen *G. xylarioides* (anamorph *F. xylarioides*), as composed by H. Maraite based on drawings from Van den Abeele and Vandenput (1956), Meiffren (1957), Booth (1971), von Blittersdorff and Kranz (1976) and Pochet (1988). (Courtesy of H. Maraite, Université Catholique de Louvain, Belgium.)

7.5. Isolation of *G. xylarioides* from Plant Tissues and Soil

G. xylarioides may be readily isolated from the stems, twigs and roots of young coffee plants and mature trees exhibiting disease symptoms. The fungus may also be grown on a range of suitable agar media. As with other fusaria, this is usually achieved by placing small pieces of surface sterilized wood excised from symptomatic plant tissues on to the surface of tap water agar medium (Booth, 1971). After a few days, a small amount of fungal material from an emerging colony may be transferred on to fresh Spezieller Nährstoffarmer agar medium (Nirenberg, 1976) to induce more prolific growth and sporulation. Isolations are most successful when made from wood pieces excised from immediately below the bark and exhibiting the blue–black discolouration typical of CWD (see 7.3). Fungal cultures should be incubated at 25°C and, if possible, under daylight fluorescent light tubes. Some exposure of cultures to ultraviolet light may improve sporulation.

In contrast, and while it can be successful, isolation of *G. xylarioides* from soil has often proven problematic. This is largely due to more rapid colonization of the agar medium by other fungi, including other fusaria (see 7.10), and is a constraint to *in vivo* research. These fungi may compete with and inhibit the normal growth of *G. xylarioides* or may simply obscure the presence of the pathogen in culture. Colonization of, and development of CWD symptoms on, susceptible coffee grown in either naturally contaminated or artificially inoculated soil is an indication of the presence of the fungus. It also highlights the potential of this approach, albeit time consuming, as a means of 'baiting' for *G. xylarioides*. More rapid and straightforward procedures for isolating the fungus or simply confirming its presence in soil (and indeed in plant material) are required, including molecular and biochemical approaches similar to those developed for fungal pathogens of other crops (Doohan *et al.*, 1998; Alves-Santos *et al.*, 2002; Knoll *et al.*, 2002).

Following isolation of the fungus from a substrate, and prior to attempting identification or more in-depth characterization, purified colonies should be prepared by transfer and culture of a single spore (microconidium or macroconidium) on fresh agar medium.

7.6. Storage and Preservation of *G. xylarioides*

Strains of *G. xylarioides* may be maintained over the short to medium term on Spezieller Nährstoffarmer agar plates or slants stored at 20°C to 25°C (for more routine use) or at 5°C (i.e. normal refrigerator temperature). For longer-term storage and preservation, strains may be preserved by depositing at very low temperatures under liquid nitrogen (cryopreservation) and/or in a freeze-dried state (lyophilization) (Ryan and Smith, 2007). Representative strains or those exhibiting characteristics of interest or value should be deposited in this manner, as should strains intended for future use. This is, of course, dependent on access to suitable facilities, something that is currently beyond the scope of many African countries. As yet, the viability and condition of strains stored for prolonged periods by cryopreservation and lyophilization has not been thoroughly evaluated. *G. xylarioides* strains, representing specific biological traits, are currently held in Genetic Resource Collections at CAB International (CABI) E-UK (Egham, UK), Centre de Coopération Internationale en Recherche Agronomique pour le Développement (CIRAD; Montpellier, France) and MUCL/BCCM (Louvain-la-Neuve, Belgium).

7.7. Cultural and Morphological Characteristics of *G. xylarioides*

As with many fusaria, the cultural and morphological characteristics exhibited *in vitro* by *G. xylarioides* depend on a number of factors, including strain, growth medium and environmental conditions, and can vary considerably. A description of the main cultural and morphological characters, sufficient to enable identification of the fungus to species level, is provided in Booth and Waterston (1964) (see below): this is based on growth of the fungus on potato sucrose agar (Booth 1971), a medium routinely used for identification of *Fusarium* species. For a full explanation of terminology, refer to *Ainsworth and Bisby's Dictionary of the Fungi* (Kirk *et al.*, 2008).

Fungal colonies are initially pale beige with sparse white mycelium. Purple discolouration may develop as the colony matures, accompanied by the production of dark bluish-black, discrete stromata, some representing ascomatal initials. Microconidia and macroconidia are produced in slimy masses on short conidiogenous cells, pionnotes or sporodochia on the vegetative mycelium. Microconidia, generally abundant, are unicellular, allantoid, curved and 5 to 10 × 2.5 to 3 µm (Plates 10 and 11). Macroconidia, which tend to be less abundant, are fusoid, falcate, 2–3 septate and 20 to 25 × 4 to 5 µm. Chlamydospores may be produced but generally are not abundant. They are oval to globose, smooth or roughened and 10 to 15 × 8 to 10 µm. Globose perithecia, 200 to 400 × 180 to 300 µm, violaceous in colour, are embedded singly or in groups in dark purple stromata. Asci within the perithecia are cylindrical, thin-walled, shortly pedicellate and 90 to 110 × 7 to 9.5 µm (Plate 12). Each ascus contains eight monostichous, hyaline to straw-coloured ascospores that are fusoid, 1–3 septate, finely roughened and 12 to 14.5 × 4.5 to 6 µm.

7.8. Variability Within *G. xylarioides*

7.8.1. Cultural and morphological variability

Cultural and morphological variability is apparent among strains of *G. xylarioides* grown on the same culture medium and maintained under uniform environmental conditions. This can be beneficial in delineating strains into intraspecific subgroups, and although these can relate to genetic traits, field behaviour and other attributes, a clear link is often not apparent. Differences, however, have been observed in pigment production (Plate 13), growth rate, growth response to temperature (optimum and maximum for growth in culture) and teleomorph form between strains associated with *C. arabica* and those associated with *C. canephora* and excelsa coffee (Girma, 2004; Tshilenge-Djim *et al.*, 2004; Lepoint, 2006), some of which reflected newly defined biological species (BS) designations (Lepoint, 2006) and *formae speciales* (Girma, 2004) (see 7.9). Although examination of such traits provides opportunities for identification and differentiation of strains, there is a need to exercise caution, given that environmental conditions alone, which vary considerably across laboratories even within East Africa, can influence the cultural characteristics of a strain significantly.

For further information on methods of isolation, culture, purification, identification, storage and preservation of *G. xylarioides* and other *Fusarium* species occurring on coffee, refer to Rutherford (2003).

7.8.2. Genetic variability and the role of the sexual cycle

The presence of a perfect state in *G. xylarioides* implies at least some degree of variability within the coffee wilt pathogen, but until recently, little was known about the underlying genetics of the fungus, the extent of genetic variability (if any) that exists or how it relates to other attributes, including field behaviour. This was partly a consequence of limited expertise and methodology to investigate genetic diversity. However, the recent re-emergence and devastating effects of the disease in Africa, coupled with development and improved access to modern molecular approaches, has renewed interest in the fungus. As a consequence, a number of research centres across the world have now completed comprehensive studies to investigate genetic variability within *G. xylarioides* and to explore the relationship between *G. xylarioides* and other fusaria (namely from coffee) from a diagnostic and taxonomic perspective.

As facilities and expertise required to undertake molecular research in Africa remain limited, the majority of such work has been undertaken at CABI, CIRAD and the Université catholique de Louvain (UCL) in the UK, France and Belgium, respectively. A broad range of primarily PCR-based approaches — e.g. analyses of microsatellite loci, presumptive mitochondrial DNA restriction fragment length polymorphisms, amplified fragment length polymorphisms, inter-simple sequence repeat [ISSR] anchored sequences, nuclear housekeeping genes [translation elongation factor, *tef*, calmodulin, CL, histone H3] and intergenic spacer and internal transcribed spacer regions — were applied to explore intraspecific genetic variability within designated regions of the *G. xylarioides* genome and within specific cell structures (e.g. nuclei, mitochondria, ribosomes). Analysis of internal transcribed spacer, for example, has been successfully used to delimit species, whereas analysis of ISSR, intergenic spacer and microsatellites can reveal interspecific and intraspecific variability. ISSR analysis in particular can reveal genetic variability within fungi in a similar manner to random amplified polymorphic DNA but is deemed to provide more reliable discrimination of individual fungal strains.

As may be expected, the various molecular approaches differed in the manner and extent to which they exposed genetic variability. Nevertheless, the findings of these studies, which examined several hundred strains of *G. xylarioides*, were generally consistent and provided an intriguing insight into genetic diversity and how it relates to other recognized traits. Although overall variability within the pathogen would appear to be limited, the research suggests that two major genetically distinct populations are responsible for current outbreaks across DRC, Uganda, Tanzania and Ethiopia (Girma, 2004; Tshilenge-Djim *et al.*, 2004; Adugna *et al.*, 2005; Lepoint, 2006; Rutherford, 2006; Bieysse, 2007). The first of these consists of *G. xylarioides* obtained from *C. canephora* affected by CWD in Uganda, DRC and Tanzania since re-emergence of the disease. Interestingly, this group also includes a single strain obtained from excelsa in Uganda. The second population consists of *G. xylarioides* obtained from *C. arabica* affected by CWD in Ethiopia. Despite the level of resolution possible with some of the molecular approaches applied, genetic diversity is not apparent within either of the two populations. As yet, therefore, it has not been possible to more precisely define the origin of strains in each population either geographically or in terms of the coffee species or cultivar of origin. This is perhaps surprising, given that production of perithecia by the fungus is frequently observed on coffee trees under suitable conditions in the field (Van der Graaff and Pieters, 1978; Girma *et al.*, 2001) and assuming, of course, that the perithecia so observed are a reliable indicator of sexual reproduction.

Genetic variability also exists among *G. xylarioides* strains recovered from *C. canephora* and *C. excelsa* during earlier CWD epidemics in CAR, Guinea and Ivory Coast, all of which differ genetically to the two populations responsible for current outbreaks. Unfortunately, few strains are available for study from these earlier outbreaks, and these may have become altered due to routine sub-culturing on nutrient-rich culture media since their isolation. Uncertainty therefore remains as to how and why this variability may have arisen and how it relates to field behaviour. Indeed, whether any of these strains were responsible for the earlier outbreaks, and to what extent, remains unknown. Furthermore, we do not know precisely how these strains relate to those currently affecting *C. canephora*, *C. excelsa* and *C. arabica* in DRC, Uganda, Tanzania and Ethiopia. It is feasible that the limited genetic diversity observed in current strains may have arisen as a consequence of (i) selection pressure imposed by widespread cultivation of 'resistant' coffee genotypes (species or varieties) to counteract the earlier epidemics and/or (ii) the relative fitness and rapid spread of a single or small number of survivors or variants, perhaps in response to cultivation of resistant germplasm. Apparent sexual incompatibility observed in the fungus (see below) may have also played a role in emergence of the genetic variants encountered to date.

The existence of a functioning sexual cycle in *G. xylarioides*, as in any organism, can play a significant role in the exchange of genetic information and hence the ability of the pathogen to adapt to changing conditions, including those brought about by modification of crop management practices. Booth (1971) considered *G. xylarioides* to be a heterothallic fungus composed of sex-linked male and female forms that could be distinguished by their morphological features. As described below, Booth's inclusion of the male form within *G. xylarioides* now appears inappropriate, as genetic studies show it to be more closely related to *Fusarium lateritium* than to *G. xylarioides* (Geiser *et al.*, 2005; Lepoint *et al.*, 2005). Booth's original description may in some ways be understandable, given that *F. lateritium* is also a recognized pathogen of coffee, and simultaneous colonization of CWD affected plants by several fusaria is observed (Tshilenge-Djim *et al.*, 2004, Lepoint 2006).

Recent research nevertheless confirms that *G. xylarioides* is indeed heterothallic. At UCL, the teleomorph of the fungus was successfully generated when two tester strains of opposing mating type, each derived from a single conidium or ascospore, were paired *in vitro* (Lepoint *et al.*, 2005). Pairing strains obtained from different geographic locations and coffee species enabled discrimination of three BS, BS1, BS2 and BS3, within which strains were sexually compatible (Plate 14). A sterility group (SG4), composed of a number of reproductively sterile strains, was also identified. The latter are sexually incompatible with each other and with strains in each of the BS, whereas strains belonging to any one BS are sexually incompatible with those of another. It should be noted that in these mating tests, teleomorph production and sexual compatibility are confirmed by production of fertile ascospores and not merely fusion of fungal mycelium or production of empty perithecia ('protothecia') (Plate 15). To date, morphological discrimination of mating types has not been possible, all strains exhibiting the characteristics previously described for Booth's female form.

In these studies, a clear relationship between the genetic and biological attributes of the pathogen was observed, in that BS1 and BS2 comprised strains associated, respectively, with *C. arabica* in Ethiopia and with *C. canephora* in DRC, Uganda and Tanzania (Lepoint, 2006). BS3 comprised the small number of strains associated with CWD during the earlier and very damaging outbreaks in Central and West Africa. Of importance, and given the time required to perform *in vitro* mating assays, molecular characterization (PCR amplification and sequencing) of the mating type gene (*MAT*), based on primer pairs previously developed for *F. oxysporum* and the *Gibberella fujikuroi* species complex (GFC), enabled the identification of mating type idiomorphs, *MAT*1-1 and *MAT*1-2, thereby supporting the hypothesis that *G. xylarioides* sensu lato is heterothallic (Lepoint *et al.*, 2005; Lepoint, 2006). This now enables mating types to be identified before confirmatory crosses are performed or where these are not successful. Taxonomically, and although there are still differing opinions as to their suitability and relevance, sequencing and phylogenetic analysis of *MAT* loci also place the fungus within the African clade of the GFC, an important taxon comprising pathogenic fusaria, and resolves four distinct phylogenetic species corresponding to the BS and SG revealed in the mating tests. These lineages were also resolved by random amplified polymorphic DNA analysis and amplification and sequencing of a combination of non-*MAT* nuclear genes (Fig. 7.2). Of note, infertility of mating crosses between *G. xylarioides* and recognized GFC mating populations suggests the coffee wilt pathogen to be a new BS within the GFC. Mating type tester strains identified for the four lineages have been deposited for secure, long-term storage at MUCL/BCCM.

The findings concerning the reproductive cycle may, at least in part, provide an explanation as to why perithecia production readily occurs in nature but (to the authors' knowledge) has never been observed on coffee plants inoculated with fungal cultures derived from a single conidiospore. The ability to cross representative strains of *G. xylarioides* in such mating tests also offers new possibilities to further explore the nature, origins and transfer

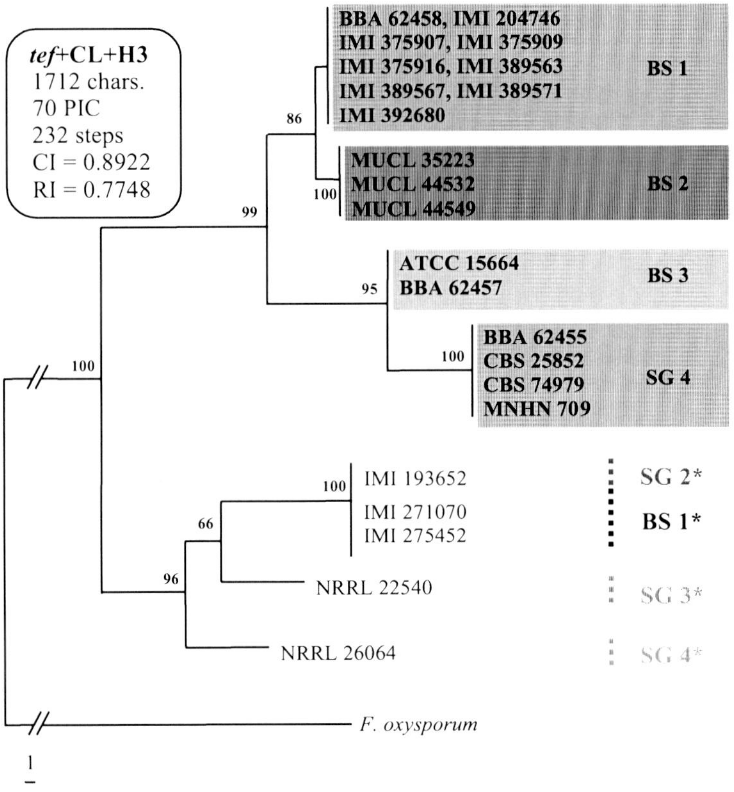

Fig. 7.2. Maximum-parsimony phylograms based on the combined autosomal (*tef* + CL + H3) data set of representative strains of the *G. xylarioides* complex (GxC) and *Gibberella indica* complex from diverse geographical and host origins. BS and SG defined in carrot agar crosses are indicated by a coloured box for the GxC and by a dotted line for the *G. indica* complex next to terminal clades resolved. Trees were generated with PAUP v.4.0 b10 (Swofford, 2001) using *F. oxysporum* as outgroup and available National Center for Biotechnology Information sequences for closely related species belonging to the GFC African clade (O'Donnell *et al.*, 1998). Bootstrap values based on 1000 replications are indicated in percentages at internodes when replication frequencies exceed 50%. (Courtesy of P. Lepoint and H. Maraite, UCL, Belgium.)

of important characteristics of the CWD pathogen and to more accurately assess the likelihood of new variants emerging and implications for future disease management.

7.8.3. Pathogenic variability

Extensive host range testing has been undertaken both under natural (e.g. screen house) environments in Africa and controlled (glasshouse) conditions in Europe to clarify the nature of the plant–pathogen relationship in CWD (Pieters and Van der Graaff, 1980; Tshilenge-Djim *et al.*, 2004; Adugna *et al.*, 2005; Girma *et al.*, 2006; Kilambo *et al.*, 2006; Musoli *et al.*, 2006; Bieysse, 2007). The primary aim of these studies was to evaluate specificity and aggressiveness in the pathogen and resistance within coffee, partly as a means of identifying possible sources of resistance for use in ongoing breeding programmes. They involved inoculation, by various means, of a range of coffee species, cultivars and clones obtained from Africa and other coffee-producing areas, including *C. canephora*, *C. arabica* and *C. excelsa*, with purified strains of *G. xylarioides* obtained from different coffee species and locations in Africa where CWD is, or was, prevalent. Strains associated with both the current outbreaks in DRC, Uganda, Tanzania and Ethiopia and with the earlier outbreaks in West and Central Africa were included in these studies.

Host specificity within the pathogen was clearly shown, with the majority of strains inducing CWD symptoms only on the coffee species from which they had been derived (Chapter 8). The findings reflect those of other approaches to pathogen characterization, including the studies of genetic diversity described above, as they again delineated *G. xylarioides* into two geographically defined populations – one comprising strains originating from Ethiopia and pathogenic towards *C. arabica*, and the other strains originating from DRC, Uganda and Tanzania and pathogenic to *C. canephora*. The findings also confirm earlier suspicions of host specificity based solely on the observed occurrence of CWD on *C. arabica* and *C. canephora* in the respective countries, including the findings of comprehensive and systematic surveys undertaken across East and Central Africa in 2002 and 2003 (Oduor *et al.*, 2004; see also Chapters 3, 4 and 5). It is also of significance that where *C. arabica* and *C. canephora* are cultivated in close proximity and even under conditions of severe disease pressure, CWD will only affect one of the two species (Kilambo *et al.*, 2006).

Of significance, the few strains associated with the earlier CWD outbreaks in Central and West Africa were also assessed for pathogenicity but do not show the same degree of host specificity. One strain in particular, isolated in the 1960s from excelsa coffee affected by CWD in the CAR, is cross-pathogenic to excelsa, *C. arabica* and also *C. canephora*.

The new knowledge acquired on genetic diversity within *G. xylarioides*, on the pathogenic nature of the fungus and on the sexual cycle has led to new species and *formae speciales* concepts being proposed, as described below.

7.9. Taxonomic Status of *G. xylarioides*

G. xylarioides was previously recognized as a member of the *Fusarium* section *Lateritium*. However, recent phylogenetic studies (including those described above) on strains that match both the original morphological description of the species (Steyaert, 1948) and of the female form as described by Booth (1971) have shown that these belong to the African clade of the GFC, a clade replacing *Fusarium* section *Liseola* (O'Donnell *et al.*, 1998; O'Donnell *et al.*, 2000; Geiser *et al.*, 2005; Lepoint *et al.*, 2005). Strains exhibiting the morphology of the 'male' strain, as described by Booth, have been found to be genetically more closely related to *F. lateritium*.

As described here and in other chapters of this book, recent research on the CWD pathogen has combined conventional approaches with some of the most up-to-date analytical methods currently available and has provided invaluable new information on diversity within the fungus and how it relates to behaviour. It is clear that a number of geographically isolated forms exist that are genetically diverse, sexually incompatible, pathogenic to particular coffee species and show at least some cultural and morphological variability. Each could represent a distinct phylogenetic (evolutionary) lineage in what was originally considered to be a single species, suggesting the existence of a species complex (GxC) composed of several cryptic species. Based on the findings, the taxonomic status of the fungus has been reviewed and to date, the following taxa proposed (Lepoint *et al.*, 2005; Lepoint, 2006):

- *G. abyssiniae* (anamorph, *F. abyssiniae*) sp. nov. – responsible for current CWD outbreaks on *C. arabica* in Ethiopia;
- *G. congoensis* (anamorph, *F. congoensis*) sp. nov. – responsible for CWD outbreaks on *C. canephora* in the DRC;
- *F. guineensis* sp. nov. – described using *C. canephora*-associated strains obtained during the earlier outbreaks in Guinea, Côte d'Ivoire and possibly CAR;
- *G. xylarioides* (*F. xylarioides*) – neotypified using Central African *C. excelsa* strain BBA 62457 (=DSMZ 62457) and strain ATCC 15664 collected during the first CWD epidemic.

The first three species are previously undescribed.

Similarly, and principally based on host–pathogen interactions and molecular characterization, two *formae speciales* have been proposed for strains affecting *C. arabica*, *C. canephora* and *C. excelsa* (Girma, 2004; Girma *et al.*, 2006) (Chapter 8). These correspond to the first two taxa described above and are, respectively:

- *G. xylarioides* f. sp. *abyssiniae* (anamorph: *F. xylarioides* f. sp. *abyssiniae*)
- *G. xylarioides* f. sp. *canephorae* (anamorph: *F. xylarioides* f. sp. *canephorae*).

Caution must be exercised when interpreting the findings of the research studies completed to date. Given the number and geographically diverse origins of strains examined and the breadth of characterization methods employed, there is good evidence for two biological populations or species being responsible for current CWD problems in East and Central Africa. Although observed pathogenic specificity to coffee suggests that these may represent two *formae speciales*, a designation that implies a strict relationship with a particular host, evaluation of strains against other potential hosts has been limited. History has shown that a number of fungi assigned to *formae speciales* have subsequently been found to have a broader host range (albeit encompassing primary and secondary hosts) than originally described. Fusarium wilt of cotton is but one example (Hillocks, 1992).

The limited number of strains available for study from the earlier epidemics also limits our ability to draw firm conclusions as to their role in CWD development and their relationship with current strains. Have, for example, these earlier strains been eradicated? Are they still present and perhaps still responsible for CWD outbreaks, albeit localized and perhaps unrecognized or not reported (on either cultivated or wild coffee)? Do other variants capable of attacking *C. arabica* or *C. canephora* already exist and simply await dispersal to, or the arrival of, a suitable, susceptible host? Further assessment of the current situation in CAR, Guinea, Ivory Coast and neighbouring countries in Central and West Africa is required to address these questions, coupled with access to a greater number of *G. xylarioides* originating from these regions.

7.10. Other Fusaria on Coffee

A number of other *Fusarium* species occur on coffee, including *Fusarium solani* (Mart.) Sacc. (*Nectria haematococca*), *Fusarium stilboides*, *F. lateritium*, *F. oxysporum* (a true vascular wilt) and *Fusarium decemcellulare* (Waller and Holderness, 1997; Rutherford, 2003; Tshilenge-Djim *et al.*, 2004; Rutherford and Phiri, 2006; Serani *et al.*, 2007; Waller *et al.*, 2007). All are recognized as pathogens of coffee in their own right, and all have been recovered from trees affected by CWD, in some cases from the same excised wood pieces from which *G. xylarioides* was isolated (Rutherford and Flood, 2005; Lepoint, 2006). The nature of the association of these fungi with *G. xylarioides* is not clear. They may, under some circumstances, coexist with *G. xylarioides* as components of a CWD 'complex' (Rutherford, 2006; Lepoint, 2006) and may therefore have some influence not only on development of what are considered to be CWD symptoms but also on crop yield and quality. Importantly, the external and internal symptoms that these fusaria induce can be very similar to those of CWD and may lead to an inaccurate diagnosis (Waller and Holderness, 1997; Tshilenge-Djim *et al.*, 2004; Serani *et al.*, 2007). *F. solani*, for example, also causes wilt symptoms accompanied by a dry root rot. Coffee bark disease, caused by *F. stilboides* Wollwenw. (*Gibberella stilboides*), is denoted by a progressive decline in host vigour. Discolouration of the wood beneath the bark may also develop in both diseases, *F. solani* causing a reddish-purple or brown discolouration in the roots and collar region for example, but this is usually unlike the blue–black staining typical of *G. xylarioides*. Tshilenge-Djim *et al.* (2004) noted that a number of fusaria originally isolated from wilted coffee could induce wilt-like symptoms when re-inoculated on to coffee plantlets. However, death of plantlets was only observed following inoculation with *G. xylarioides*.

Invasion of coffee by several fusaria may obscure symptoms, external and internal, that would otherwise be readily visible and allow a clear diagnosis. This highlights the need for sufficient awareness of the possible occurrence of these various fusaria on coffee and of symptoms characteristic for each. It also stresses the need to confirm the presence, identity and pathogenic nature of an apparent causal organism using appropriate methods, including those outlined here and in other chapters of this book. *G. xylarioides*, *F. solani*, *F. stilboides* and the other fusaria highlighted here are distinguishable by the cultural (e.g. colony pigmentation) and morphological characteristics they exhibit when grown on an appropriate agar medium such as potato sucrose agar (Waller and Holderness, 1997; Rutherford, 2003). Generally, *G. xylarioides* is readily distinguishable from these other species by its curled or crescent-shaped microconidia. However, and although not normally associated with coffee, care should be taken not to confuse microconidia of *G. xylarioides* with *G. indica* (anamorph *Fusarium udum*), a known pathogen of pigeon pea (*Cajanus cajan*) (Hillocks and Songa, 1993). Although examination of morphological characters remains a fundamental requirement for species identification, molecular tools and other approaches to identification and characterization of *G. xylarioides* and other fusaria associated with CWD are now available. Some have the added benefit of discriminating within, as well as between, the various fungal species.

7.11. Conclusion

Recent studies on CWD, largely completed as part of the RCWP, have greatly enhanced our understanding of the occurrence, form and behaviour of the CWD pathogen, *G. xylarioides*, including from a historical perspective. It has provided important information on coffee production areas already affected by the disease and those under threat. A number of genetically and biologically distinct forms, proposed as distinct species, have been shown to be responsible for current epidemics across East and Central Africa. These are geographically delineated and exhibit pathogenic specificity to one or other of the two coffee species of economic importance in the region, *C. arabica* and *C. canephora*. Although access to forms of the pathogen associated with earlier epidemics in West and Central Africa has been limited, characterization and comparison of these with those prevalent today have provided some clarification as to their importance to current epidemics and how these may have arisen. This new knowledge, coupled with a global search to obtain coffee germplasm with resistance to CWD and the outcome of research to clarify the underlying genetic basis of resistance, provides a basis for redefining our approach to managing CWD through both prevention and eradication. Swift action is required. CWD has already devastated coffee production across large parts of Africa, destroying the livelihoods of millions of families already facing hardship, and continues to sweep across the continent. In parallel, awareness within the coffee community of the disease and how it can be tackled must be increased, particularly in areas not yet suffering but under imminent threat. Research efforts must also continue to allow us to build upon what we have already learned and to seek an effective and sustainable means to eradicate CWD or, in the short term, maintain disease levels within an acceptable economic threshold.

Acknowledgements

The authors would like to thank those who provided information required for the preparation of this chapter, in particular colleagues and collaborators from: CABI, UK and Kenya; UCL, Belgium; CIRAD, France; Coffee Research Institute, Uganda; Ethiopian Agricultural Research Organisation; Tanzania Coffee Research Institute and Université de Kinshasa, DRC. Coordinated research on CWD in Africa and Europe was performed in the framework of European Community-funded project International Cooperation-COWIDI ICA4-CT-2001-10006, coordinated by CIRAD, and UK Department for International Development Crop Protection Programme-funded project ZA0505, coordinated by CABI E-UK. The Common Fund for Commodities is also acknowledged for providing financial support to the overall RCWP. The authors would also like to acknowledge the considerable efforts made by research scientists, extension services, farmers and other coffee stakeholders in Africa and elsewhere who completed the many challenging studies of coffee wilt to help improve our understanding of the disease. Without their contribution, the new knowledge presented here and in other chapters of this book would not have become available. Thanks also go to Dr J.M. Waller (CABI Emeritus Fellow) and Dr J. Flood (CABI E-UK) for kindly reviewing and providing advice on this chapter.

References

Adugna, G., Hindorf, H., Steiner, U., Nirenberg, H.I., Dehne, H.W. and Schellander, K. (2005) Genetic diversity in the coffee wilt pathogen (*Gibberella xylarioides*) populations: differentiation by host specialization and RAPD analysis. *Zeitschrift fur Pflanzenkrankheiten und Pflanzenschutz* 112, 134–145.

Alves-Santos, F.M., Ramos, B., García-Sánchez, M.A., Eslava, A.P. and Díaz-Mínguez, J.M. (2002) A DNA-based procedure for in planta detection of *Fusarium oxysporum* f. sp. *phaseoli*. *Phytopathology* 92, 237–244.

Beckman, C.H. (1987) *The Nature of Wilt Diseases of Plants*. The American Phytopathology Society, St. Paul, MN.

Bieysse, D. (2005) (ed.) *Development of a Long-Term Strategy Based on Genetic Resistance and Agroecological Approaches Against Coffee Wilt Disease in Africa: Fourth Annual Report. INCO-DEV Contract ICA4-CT-2001-10006*. CIRAD-AMIS, Montpellier, France.

Bieysse, D. (2006) (ed.) *Development of a Long-Term Strategy Based on Genetic Resistance and Agroecological Approaches Against Coffee Wilt Disease in Africa: Fifth*

Annual Report. INCO-DEV Contract ICA4-CT-2001-10006. CIRAD-AMIS, Montpellier, France.

Bieysse, D. (2007) (ed.) *Development of a Long-Term Strategy Based on Genetic Resistance and Agroecological Approaches Against Coffee Wilt Disease in Africa: Final Report. INCO-DEV Contract ICA4-CT-2001-10006*. CIRAD-AMIS, Montpellier, France.

Booth, C. (1971) *The Genus Fusarium*. Commonwealth Mycological Institute, Kew, UK.

Booth, C. and Waterston, W.M. (1964) *CMI Descriptions of Plant Pathogenic Fungi and Bacteria No. 24*. CAB International, Wallingford, UK.

Doohan, F.M., Parry, D.W., Jenkinson, P. and Nicholson, P. (1998) The use of species-specific PCR-based assays to analyse *Fusarium* ear blight of wheat. *Plant Pathology* 47, 197–205.

Fraselle, J. (1950) Observations préliminaires sur une trachéomycose de *Coffea robusta*. *Bulletin Agricole du Congo Belge* XLI, 361–372.

Figueres, R. (1940) Sur une maladie très grave du caféier en Oubangui. Rapport. Ministère des Colonies, Paris, France.

Flood, J. and Brayford, D. (1997) Current knowledge of pathogen variability, epidemiology and control of *Fusarium xylarioides*. In: *1st Regional Coffee Wilt Workshop*, Kampala, Uganda.

Geiser, D.M., del Mar Jiménez-Gasco, M., Kang, S., Makalowska, I., Veeraraghavan, N., Ward, T.J., Zhang, N., Kuldau, G.A. and O'Donnell, K. (2004) FUSARIUM-ID v. 1.0: a DNA sequence database for identifying *Fusarium*. *European Journal of Plant Pathology* 110, 473–479.

Geiser, D.M., Lewis Ivey, M.L., Hakiza, G., Juba, J.H., and Miller, S.A. (2005) *Gibberella xylarioides* (anamorph: *Fusarium xylarioides*), a causative agent of coffee wilt disease in Africa, is a previously unrecognized member of the *G. fujikuroi* species complex. *Mycologia* 97, 191–201.

Girma, A.S. (2004) Diversity in Pathogenicity and Genetics of Gibberella xylarioides (Fusarium xylarioides) Populations and Resistance of *Coffea* spp. in Ethiopia. PhD dissertation. University of Bonn, Bonn, Germany.

Girma, A.S., Flood, J., Hindorf, H., Bieysse, D., Simons, S. and Rutherford, M. (2006) Tracheomycosis (*Gibberella xylarioides*)—a menace to world coffee production: evidenced by cross inoculation of historic and current strains of the pathogen. In: *Proceedings of the 21st International Scientific Conference on Coffee Science (ASIC)*. Montpellier, France, pp. 1268–1276.

Girma, A., Mengistu, H. and Hindorf, H. (2001) Incidence of tracheomycosis, *Gibberella xylarioides* (*Fusarium xylarioides*), on arabica coffee in Ethiopia. *Zeitschrift fuÉr Pflanzenkrankheiten und Pflanzenschutz* 108, 136–142.

Hillocks, R.J. (1992) Fusarium wilt. In: Hillocks R.J. (ed.) *Cotton Diseases*. CAB International, Wallingford, UK, pp. 127–160.

Hillocks, R.J. and Songa, W. (1993) Root-knot and other nematodes associated with pigeonpea plants infected with *Fusarium udum* in Kenya. *Afro-Asian Journal of Nematology* 3, 143–147.

Jacques-Felix, H. (1950) Première action contre la trachéomycose du caféier en Côte d'Ivoire. *Agronomia Tropical Nogent*, 12 pp.

Jacques-Felix, H. (1954). La carbunculariose. *Bulletin Scientifique—Centre de Recherches Agronomiques (Francia)* 5, 296–344.

Kangire, A., Kytere, D., Hakiza, G., Warren, H., Erbaugh, M. and Kabole, C. (2002) Determination of alternate hosts for *F. xylarioides*, the causal agent of Coffee Wilt Disease (CWD) in Uganda. Internal Report. Coffee Research Institute, Uganda.

Kilambo, D.L., Kaiza, D.A. and Swai, F.B. (1997) Observation of wilting disease in some coffee growing areas in Tanzania. In: *Proceedings of the 1st Regional Workshop on Coffee Wilt /Disease (Tracheomycosis).* International Conference Centre, Kampala, Uganda, pp. 61–65.

Kilambo, D.L., Ng'homa, N.M., Mohamed, R.A., Teri, J.M., Poole, J., Flori, A. and Pinard, F. (2004) Coffee disease survey in Tanzania. In: *Proceedings of the 20th International Conference on Coffee Science (ASIC).* Bangalore, India, pp. 1263–1266.

Kilambo, D.L., Ng'homa, N.M., Mtenga, D.J., Teri, J.M., Nzallawahe, T., Rutherford, M. and Masumbuko, L. (2006) Progress towards searching for durable resistance to *Fusarium* wilt (*Fusarium xylarioides*) in *Coffea canephora* germplasm in Tanzania. In: *Proceedings of the 21st International Scientific Conference on Coffee Science (ASIC).* Montpellier, France, pp. 1386–1389.

Kirk, P.M., Cannon, P.F., Minter, D.W. and Stalpers, J.A. (2008). *Dictionary of the Fungi,* 10th edn. CABI Publishing, Wallingford, UK.

Knoll, S., Mulfinger, S., Niessen, L. and Vogel, R.F. (2002) Rapid preparation of *Fusarium* DNA from cereals for diagnostic PCR using sonification and an extraction kit. *Plant Pathology* 51, 728–734.

Kranz, J. (1962) Coffee disease in Guinea. *FAO Plant Protection Bulletin* 10, 107–109.

Kranz, J. and Mogk, M. (1973) *Gibberella xylarioides* Heim et Saccas on arabica coffee in Ethiopia. *Phytopathologische Zeitschrift* 78, 365–366.

Lepoint, P.C.E. (2006) Speciation within the African coffee wilt Pathogen. PhD thesis. Université catholique de Louvain, Louvain, Belgium.

Lepoint, P.C.E., Munaut, F.T.J. and Maraite, H.M.M. (2005) *Gibberella xylarioides* sensu lato from *Coffea canephora*: a new mating population in the *Gibberella fujikuroi* species complex. *Applied and Environmental Microbiology* 71, 8466–8471.

Meiffren, M. (1957) La trachéomycose. Les maladies du caféier en Côte d'Ivoire. Haut Commissariat de l'A.O.F., Centre de Recherches Agronomiques de Bingerville, Côte d'Ivoire. *Bulletin de la recherche agronomique du Bingerville* 13, 9–54.

Meiffren, M. (1961) Contribution aux recherches du caféier en Côte d'Ivoire. *Café Cacao Thé* 5, 28–37.

Musoli, P.C. (2007). Recherche de sources de résistance à la trachéomycose du caféier canephora Pierre, due à *Fusarium xylarioides* en Ouganda. Thesis. Université de Montpellier II, Montpellier, France.

Musoli, P.C., Aluka, P., Cubry, P., Dufour, M., de Bellis, F., Nakendo, S., Nabaggala, A., Ogwang, J., Kyetere, D., Leroy, T., Bieysse D. and Charrier, A. (2006) Fighting against coffee wilt disease: Uganda wild *C. canephora* genetic diversity and its usefulness. In: *Proceedings of the 21st International Scientific Conference on Coffee Science (ASIC).* Montpellier, France, pp. 1268–1276.

Musoli, P.C. and Hakiza, G. (2007) Individual final report, National Agricultural Research Organisation (NARO). In: Bieysse, D. (ed.) *Development of a Long-Term Strategy Based on Genetic Resistance and Agroecological Approaches against Coffee Wilt Disease in Africa: Final Report. INCO-DEV Contract ICA4-CT-2001-10006.* CIRAD-AMIS, Montpellier, France, pp. 56–70.

Musoli, P.C., Pinard, F., Charrier, A., Kangire, A., ten Hoopen, G.M., Kabole, C., Ogwang, J., Bieysse, D. and Cilas, C. (2008) Spatial and temporal analysis of coffee wilt disease caused by *Fusarium xylarioides* in *Coffea canephora. European Journal of Plant Pathology* (in press).

Nirenberg, H.I. (1976) Untersuchungen über die morphologische und biologische Differenzierung in der *Fusarium* Sektion *Liseola. Mitteilungen aus der Biologischen Bundesanstalt fuÉr Land- und Forstwirtschaft* (Berlin-Dahlem) 169, 1–117.

O'Donnell, K., Cigelnik, E. and Nirenberg, H.I. (1998) Molecular systematics and phylogeography of the *Gibberella fujikuroi* species complex. *Mycologia* 90, 465–493.

O'Donnell, K., Nirenberg, H.I., Aoki, T. and Cigelnik, E. (2000) A multigene phylogeny of the *Gibberella fujikori* species complex: detection of additional phylogenetically distinct species. *Mycoscience* 41, 61–78.

Oduor, G., Phiri, N., Hakiza, G.J., Abebe, M., Asiimwe, T., Kilambo, D.L., Kalonji-Mbuyi, A., Pinard, F., Simons, S., Nyasse, S. and Kebe, I. (2004) Surveys to establish the spread of coffee wilt disease, *Fusarium* (*Gibberella*) *xylarioides*, in Africa. In: *Proceedings of the 20th International Scientific Conference on Coffee Science (ASIC).* Bangalore, India, pp. 1252–1255.

Onesirosan, P.T. and Fatunla, T. (1976) Fungal fruit rots of tomatoes in Southern Nigeria. *Journal of Horticultural Science* 51, 473–479.

Pegg, G.F. (1981) Biochemistry and physiology of pathogenesis. In: Mace, M.E., Bell, A.A. and Beckman, C.H. (eds.) *Fungal Wilt Diseases of Plants.* Academic Press, New York and London, pp. 193–252.

Pieters, R. and Van der Graaff, N.A. (1980) Resistance to *Gibberella xylarioides* in *Coffea arabica*: evaluation of screening methods and evidence for the horizontal nature of the resistance. *Netherlands Journal of Plant Pathology* 86, 37–43.

Pizzinatto, M.A. and Menten, J.O.M. (1991) Pathogenicity of eight *Fusarium* species isolated from seeds to cotton seedlings. *Summa Phytopathologica* 17, 124–134.

Pochet, P. (1988) La trachéomycose du caféier robusta. *Publications Agricoles* 19, 28 pp.

Rutherford, M. (2003) *Isolation and Identification of Fusaria from Coffee. A Training Manual.* CABI International, Egham, UK.

Rutherford, M. (2006) Current knowledge of coffee wilt disease, a major constraint to coffee production in Africa. *Phytopathology* 96, 663–666.

Rutherford, M.A. and Flood, J. (2005) Epidemiology and variability of *Gibberella xylarioides*, the coffee wilt pathogen. *UK Department for International Development (DFID) Crop Protection Programme Project R8188: Final Technical Report.* CABI International, Egham, UK.
http://www.research4development.info/SearchResearchDatabase.asp?ProjectID=3636

Rutherford, M.A. and Phiri, N. (2006) *Pests and Diseases of Coffee in Eastern Africa: A Technical and Advisory Manual.* CAB International, Egham, UK.
http://www.research4development.info/PDF/Outputs/CropProtection/U3071 Coffee Manual.pdf

Ryan, M.J. and Smith, D. (2007) Cryopreservation and freeze-drying of fungi employing centrifugal and shelf freeze-drying. In: Day, J.G. and Stacey G.N (eds.) *Cryopreservation and Freeze Drying Protocols*, 2nd edn. Hna Press, New Jersey, pp. 127–140.

Saccas, A.M. (1951) La trachéomycose (carbunculariose) des *Coffea excelsa*, neo-arnoldiana et robusta en Oubangui-Chari. *Agronomie Tropicale* 6, 453–506.

Serani, S. (2000) An investigation of Fusarium spp. associated with coffee and other plants as potential pathogens of robusta coffee. MSc dissertation. Makerere University, Kampala, Uganda.

Serani, S., Taligoola, H.K. and Hakiza, G.J. (2007) An investigation into *Fusarium* spp. associated with coffee and banana plants as potential pathogens of robusta coffee. *African Journal of Ecology*, 45, 91–95.

Swofford, D.L. (2001) PAUP* Phylogenetic Analysis Using Parsimony (*and other methods), Version 4. Sinauer Associates, Sunderland, Massachusetts.

Stewart, R.B. (1957) Some diseases occurring in Kaffa Province, Ethiopia. Imperial Ethiopian College of Agriculture and Mechanical Arts, Alemaya, Ethiopia, pp. 15–16.

Steyaert, R.L. (1948) Contribution à l'étude des parasites des végétaux du Congo Belge. *Bulletin de la Societe Royale de Botanique de Belgique* 80, 11–58.

Tshilenge-Djim, P., Munaut, F., Kalonji-Mbuyi, A. and Maraite, H. (2004) Caractérisation des *Fusarium* spp. associées au dépérissement du caféier robusta en République Démocratique du Congo. *Parasitica* 60, 19–34.

Van Den Abeele, M. and Vandenput, R. (1956) *Les principales cultures du Congo Belge*, 3rd edn. Royaume de Belgique, Ministère des colonies, Publication de la Direction de l'agriculture, des forêts et de l'élevage, Bruxelles, Belgium. 932 pp.

Van der Graaff, N.A. and Pieters, R. (1978) Resistance levels in *Coffea arabica* to *Gibberella xylarioides* and distribution patterns of the disease. *Netherlands Journal of Plant Pathology* 84, 117–120.

von Blittersdorff, R. and Kranz, J. (1976) Vergleichenden.Untersuchundgen an *Fusarium xylarioides (Gibberella xylarioides Heim & Saccas)*, dem Erreger der Tracheomykose des Kaffes. *Zeitschrift fur Pflanzenkrankheiten und Pflanzenschutz* 83, 529–544.

Waller, J.M., Bigger, M. and Hillocks, R.J. (2007) *Coffee Pests, Diseases and their Management.* CAB International, Wallingford, UK.

Waller, J.M. and Holderness, M. (1997) Fusarium diseases of coffee. In: *Proceedings of the First Regional Workshop on the Coffee Wilt Disease (Tracheomycosis).* Kampala, Uganda, pp. 27–36.

8 Host–Pathogen Interactions in *Coffea–Gibberella xylarioides* Pathosystem

A. Girma,[1] D. Bieysse[2] and P. Musoli[3]
[1] *Jimma Agricultural Research Center, Ethiopian Institute of Agricultural Research, PO Box 192, Jimma, Ethiopia*
e-mail: girma.adugna@yahoo.com
[2] *Centre de Coopération Internationale en Recherche Agronomique pour le Développement (CIRAD), TA41/K, Campus International Baillarguet, 34398 Montpellier, Cedex 5, France*
[3] *Coffee Research Centre, PO Box 185, Mukono, Uganda*

8.1. Introduction

Gibberella xylarioides Heim & Saccas (anamorph: *Fusarium xylarioides* Steyaert) causes a vascular wilt disease of coffee referred to as tracheomycosis. The disease caused large-scale damage to plantations of various *Coffea* spp. in Africa in the early 1950s (Wellman, 1961; Kranz, 1962; Muller, 1997). Coffee wilt disease (CWD) has been observed only on *Coffea canephora* (robusta coffee) in the Democratic Republic of Congo (DR Congo), Uganda and Tanzania and on *Coffea arabica* (arabica coffee) in Ethiopia (Girma *et al.*, 2001; Oduor *et al.*, 2003), suggesting some form of host specialization in the pathogen populations. Tracheomycosis re-emerged as a serious disease of coffee in Africa, and certain factors (shift in the host and/or in the pathogen populations, acting either independently or in concert) have been postulated about its reappearance and becoming a major constraint to coffee production in this continent (Flood and Brayford, 1997; Girma and Hindorf, 2001; Girma, 2004).

Management of soil-borne disease is a continual challenge to growers, and one of the major reasons for the limited success is our limited knowledge of the genetic structure of the pathogen populations (Watson, 1970; McDonald, 1997). According to Leung *et al.* (1993), population structure refers to the amount of genetic variation among individuals in a population, the ways in which this variation is partitioned in time and space and the phylogenetic relationships among individuals within and between subpopulations.

The most widely used method for characterizing pathogen populations is the determination of the virulence spectrum on a set of differential varieties carrying different resistance factors, and this measure of genetic variability provides information on the pathotype structure of the populations (Leung *et al.*, 1993).

Understanding pathogen population structure can contribute to improved disease management by allowing resistance genes and plant genotypes to be identified and characterized relative to the spectrum of the pathogen population and by providing a knowledge base important to the design of strategies for variety development (Leung *et al.*, 1993). Resistance to wilt diseases depends in part on genetic potential for virulence within the pathogen populations and the inoculum concentration. The occurrence of distinctly different levels of virulence in the wilt fungus indicates that certain mechanisms of resistance are not effective against the highly virulent strains, even though they are effective against most strains of the fungal species (Beckman, 1987). Resistance of a plant (or tissue) changes sequentially during growth and development; thus, certain growth stages are more favourable than others for comparison of resistant and susceptible cultivars (Beckman, 1987). According to Beckman and Talboys (1981), two sets of events have to take place before a vascular wilt disease can develop: the pathogen must gain access to the xylem of the host, and it must continue to colonize the xylem more or less extensively. However, the fact that a particular mechanism contributes to disease resistance does not necessarily mean that it is implicated in the determination of specificity in host–parasite interaction (Beckman, 1987; Beckman and Talboys, 1981).

Interaction between coffee and *G. xylarioides* (*F. xylarioides*) has been thoroughly studied through several artificial seedling inoculation tests. These studies were exclusively conducted in well-designed experiments in the greenhouse or growth room where disease parameters such as infected and dead seedlings and number of days to first symptom appearance after inoculation (incubation periods) were documented and then subjected to statistical analysis and biometrical interpretations of the results. The host–pathogen combination genetic analyses were further backed by molecular studies, fungal morphology and *in vitro* mating test (Chapter 7). Thus, this chapter elucidates the current knowledge on host–pathogen interactions in the *Coffea*-vs.-*G. xylarioides* pathosystem gained through repeated cross-inoculation studies (pathogenicity tests), supported by some preliminary histopathological observations. This chapter also includes appraisal on seedling inoculation protocols, which are the basis for testing and verifying host resistance that can be used as a standard procedure for mass screening and breeding programs.

8.2. Appraisal of Seedling Inoculation Protocols for Screening for Resistance and Studies of the Host–Pathogen Interaction

The study of host–pathogen associations involves a three-dimensional interaction between host varieties, pathogen strains and environmental variables that can affect disease expression emanating from compatible/incompatible combinations. To be able to limit the effect of these factors on host–pathogen interactions, artificial inoculation tests have been developed. Artificial inoculation allows more genotypes/cultivars to be tested. Besides, reliance on screening for resistance in the field (based on natural inoculum) is unsatisfactory due to irregular occurrence of the pathogen in time and space, lack of understanding of which strains or races are present and unfavourable conditions that may interfere with proper appraisal of host resistance. Yet, breeding programs with a perennial crop are long term, and at least 7 to 10 years is needed to select material on the basis of natural infection in field trials; thus, alternative, more rapid tests have to be investigated to help speed up screening of germplasm (Flood, 2006). Therefore, developing standard artificial screening protocols that discriminate between resistant and susceptible genotypes is an imperative and is very relevant for such soil-borne vascular diseases as CWD.

In this particular host–pathogen interaction, many different inoculations techniques are and have been employed in various laboratories in different countries over many years. Stem-nicking method of young coffee seedlings with inoculum suspension (2–2.5 × 10^6 concentration) of *G. xylarioides* isolate at cotyledon stage (2 to 2.5 months old) using a scalpel has been adopted as the preferred standard practice on *C. arabica* in Ethiopia (Pieters and Van der Graaff, 1980; Girma and Mengistu, 2000; Girma *et al.*, 2005, 2007), whereas syringe injection of inoculum (1 × 10^6 concentration) into the stem of growing seedlings (at 9 to 10 months old) has been routinely used at Centre de Coopération Internationale en Recherche Agronomique pour le Développement (CIRAD; D. Bieysse, personal communication, 2003) and the root dip inoculation procedure is employed in screening *C. canephora* seedlings (5 to 6 months old) in Uganda, DR Congo and Tanzania (Hakiza *et al.*, 2004; Kilambo *et al.*, 2007; Chapter 10).

The stem injection method is labourious and time consuming, necessitating the piercing of the stem with a syringe and deposition of the required inoculum concentration. This may have the disadvantage of disease escape and is inconvenient to screen a large number of coffee genotypes. The stem-nicking method, although allowing screening of many genotypes in a short period, does have the disadvantage of sometimes causing seedling damage. Further problems associated with artificial screening techniques include differences in resistance response (symptom expression) among coffee types due to growth stage of seedlings, which in turn relate to appearance and development of the primary and secondary xylem tissues of the plant. In addition, inoculum concentration and pre- and post-inoculation conditions (mainly temperature, relative humidity and optimum moisture of the media) are also known to influence infection and disease development (Girma and Mengistu, 2000). Thus, standardizing the inoculation protocols (methodologies) so that they can be uniformly applied with reproducible results across different laboratories/greenhouses, identifying proper growth stages of the host that show differential reactions, selection of aggressive strain/isolate and conditions that favour infection and wilt disease development is of paramount importance in designing an effective screening and breeding programme for CWD management.

To optimize the protocol, a comprehensive investigation was carried out on three *C. arabica* cultivars (designated resistant, intermediate and susceptible) from known field observations, which were inoculated with a standard arabica isolate (Gx2) (IMI 71975) at four different growth stages of seedlings (between 'soldier' and 'two pairs of true leaves') employing five inoculation methods, namely, stem nicking, stem injection, root cut and dip, root wound and transplant into artificially and naturally infested soil. The fungal inoculum was uniformly adjusted to 2×10^6 spore per millilitre for the treatment combinations. There were highly significant differences between coffee cultivars, inoculation methods and growth stages of the seedlings among the three variables (Tables 8.1 and 8.2). As indicated in Table 8.1, cotyledon and one pair of true leaf stages showed significantly higher seedling death, whereas seedlings at 'soldier stage' failed to open cotyledon leaves, which may be due to physiological shock and did not reveal typical wilting although the seedlings eventually died.

When means of inoculation methods were compared, significantly higher percentage deaths were recorded in stem nicking followed by the root dip and stem injection methods and revealed differences among coffee cultivars in their response to the disease (Table 8.2). Seedlings inoculated by the root dip method developed symptoms very slowly and sometimes failed to recover after transplanting and did not show typical CWD symptoms. The stem injection method was technically inefficient in screening large number of accessions, as it is difficult to pierce the seedling stems and place the

Table 8.1. Percentage seedling death[a] in three *C. arabica* cultivars inoculated with *G. xylarioides* isolates at different growth stages in the greenhouse[++].

	Growth stages[c]				
Coffee cultivar[b]	Soldier	Cotyledon	One pair of true leaves	Two pairs of true leaves	Mean
200/71	24.5 fg	46.3 cd	48.6 c	21.6 g	35.3 A
21/79	32.5 ef	45.2 cd	48.1 c	35.8 e	40.4 A
20/85	48.1 c	68.3 a	58.8 b	38.9 de	53.5 B
Mean	35.1 X	53.3 Y	51.8 Y	32.1 X	43.1

[a]Percentage death is calculated from cumulative number of dead plants over total number of seedlings (20 per treatment) 6 months after inoculation, and the actual wilt values were arcsine-square root transformed to normalize the data.
[b]Coffee cultivars 21/79, 201/71 and 20/85 had resistance, moderate resistance and susceptible reactions under field conditions, respectively.
[c]Growth stages were 'soldier or hypocotyls', when the seeds emerge from the soil before opening of the cotyledon; 'cotyledon', fully opened butterfly; 'one pair of true leaves', when 1st pair true leaves are fully opened; 'two pairs of true leaves', when two pairs of true leaves were fully opened.
[++]Means followed with the same letter(s) are not significantly different from each other at $P = 0.05$. Least significant difference values for mean comparison for cultivars, stages and interactions are 6.93, 4.61 and 7.99, respectively.

Table 8.2. Percentage seedling death[a] in three *C. arabica* cultivars inoculated with *G. xylarioides* isolates with different inoculation methods greenhouse[c].

	Inoculation methods[1]					
Coffee cultivars[b]	Stem nicking	Stem injection	Root dip	Artificially infested soil	Naturally infested soil	Mean
200/71	53.1 d–f	45.0 f	46.9 ef	24.0 g	7.3 h	35.3 A
21/79	70.5 ab	55.5 c–e	57.8 cd	17.0 g	1.2 h	40.4 A
20/85	61.1 cd	62.1 bc	71.5 a	47.2 ef	25.7 g	53.5 B
Mean	61.6 w	54.2 x	58.7 wx	29.4 y	11.4 z	43.1

[a]As Table 8.1.
[b]Coffee cultivars (as Table 8.1).
[c]Means followed with the same letter(s) are not significantly different from each other at $P = 0.05$. Least significant difference values for mean comparison for cultivars, methods and interactions are 6.93, 5.61 and 8.93, respectively.
[1]Inoculation methods were as described in the text.

required amount of inoculum at the later growth stages. Overall, the mean percentage seedling death in cultivar 200/71 (resistant) and cultivar 21/79 (moderately resistant) were significantly different from the susceptible cultivar (20/85) with seedling deaths of 35.3%, 40.4% and 53.5% (Tables 8.1 and 8.2).

The stem-nicking method at the cotyledon stage using spore concentration of 2×10^6 was adopted as a standard protocol for screening coffee germplasm and studying the host–pathogen interaction. Inoculated seedlings were maintained at high humidity (>95%) and temperature of about 23°C for 1 week as optimum conditions for infection. After such a post-inoculation period, the seedlings were transferred into a glasshouse for further disease development. The seedlings were not uprooted and transplanted, as this practice disturbs the plant system and may predispose them to the disease.

8.3. *C. arabica–G. xylarioides* Isolates Interactions (Arabica Cultivars vs. Isolates)

These studies involved inoculating seedlings of nine *C. arabica* cultivars that had been observed to possess various levels of resistance to the disease under field conditions with four *G. xylarioides* isolates obtained from infected *C. arabica* trees (Gx12 [IMI 375906], Gx26 [IMI 375907], Gx31 [IMI 375908] and Gx43 [IMI 375909]). The result showed highly significant ($P < 0.01$) differences among coffee cultivars and *G. xylarioides* isolates and a significant ($P < 0.05$) cultivar–isolate interaction both in terms of percentage seedling death and incubation period (from inoculation to first symptoms) (Girma and Mengistu, 2000; Girma and Hindorf, 2001). According to Van der Plank

(1984), highly significant differences among the main effects indicate the existence of horizontal resistance in the host and variation in aggressiveness in the fungus population, but a significant interaction between the cultivars and the isolates (i.e. a differential effect) in both disease parameters implies vertical resistance in the host and virulence in the pathogen (Girma and Mengistu, 2000; Girma and Hindorf, 2001).

When comparing all cultivars, 61/85, 24/85 and F-17 showed significantly ($P < 0.05$) higher disease levels with 62.6%, 60.5% and 51.4% seedling death, respectively (Table 8.3), with a shorter incubation period of about 30–60 days (Table 8.4). In contrast, a significantly ($P < 0.05$) lower percentage of seedling death (28.8%, 24.3%, and 12.0%) was observed on cultivars 35/85, 74165 and 7440, respectively (Table 8.3), with longer incubation periods ranging between 84 and 112 days (Table 8.4). Isolates Gx26, Gx43 and Gx31 caused more seedling death (58.2%, 53.4% and 52.2%, respectively) than isolate Gx12, which induced no symptoms throughout the trial (Table 8.3). The Teppi isolate (Gx26) induced wilting symptoms in a significantly shorter incubation period (82 days) as compared to Jimma isolate (Gx31), which induced symptoms in around 100 days (Table 8.4). Comparing the combined effect of cultivar vs. isolate interactions (Table 8.3), Gx26 (Teppi isolate) induced a higher rate of death on cultivars SN-5 (70.8%), 74304 (64.6%) and

Table 8.3. Percentage seedling death[a] in *C. arabica* cultivars inoculated with *G. xylarioides* isolates collected from representative fields in south-west Ethiopia.

Coffee cultivars[b]	*G. xylarioides* isolates[c]				
	Gx12	Gx26	Gx31	Gx43	Mean
74165 (R)	0.0 j	40.5 e–i	33.9 f–i	22.6 g–j	24.3 E
7440 (MR)	0.0 j[d]	17.1 h–j	11.6 i–j	19.3 h–j	12.0 F
74304 (S)	0.0 j	64.6 a–f	48.8 b–h	38.0 f–i	37.8 CD
F-17 (R)	0.0 j	77.8 a–c	52.6 a–g	75.1 a–d	51.4 AB
F-61 (MR)	0.0 j	54.8 a–g	57.1 a–f	70.8 a–e	45.7 BC
SN-5 (S)	0.0 j	70.8 a–e	62.4 a–f	46.8 c–h	45.0 BC
35/85 (R)	0.0 j	43.8 d–h	35.3 f–i	36.0 f–i	28.8 DE
61/85 (MR)	0.0 j	80.4 ab	85.1 a	85.1 a	62.6 A
24/85 (S)	0.0 j	73.8 a–d	83.0 a	85.2 a	60.5 A
Mean	0.0 N	58.2 M	52.2 M	53.4 M	

[a]Percentage seedling deaths were transformed to angular values to normalize the data before analysis. Means (column, row) followed with the same letter(s) are not significantly ($P < 0.05$) different from each other according to Duncan's multiple range test. Least significant difference values for the cultivars, the isolates and the interactions comparisons were 10.8, 9.2, and 27.6, respectively.
[b]Cultivars 74165, F-17 and 35/85 were resistant (R); 7440, F-61 and 61/85 were moderately resistant (MR); 74304, SN-5 and 24/85 were susceptible (S) to CWD under field conditions.
[c]Gx12, Gx26, Gx31 and Gx43 were *G. xylarioides* isolates obtained from Bebeka, Teppi, Jimma and Gera, respectively.
[d]0.0 indicates no external symptoms were observed during the trial.

Table 8.4. Incubation periods (in days) for *C. arabica* seedlings inoculated with *G. xylarioides* isolates collected from representative fields in south-west Ethiopia.

Coffee cultivars[a]	*G. xylarioides* isolates[b]				
	Gx12	Gx26	Gx31	Gx43	Mean
74165 (R)	0.0 h[c]	126 a–c	154 ab	154 ab	108.5 A
7440 (MR)	0.0 h	133 a–c	168 a	147 ab	112.0 A
74304 (S)	0.0 h	35 gh	112 a–e	119 a–d	66.5 CD
F-17 (R)	0.0 h	84 c–g	98 b–f	63 d–g	61.2 CD
F-61 (MR)	0.0 h	56 e–h	77 c–g	84 c–g	54.2 D
SN-5 (S)	0.0 h	98 b–f	77 c–g	105 b–f	70.0 C
35/85 (R)	0.0.h	105 b–f	133 a–c	98 b–f	84.0 B
61/85 (MR)	0.0 h	56 e–h	35 gh	28 gh	29.7 E
24/85 (S)	0.0 h	49 f–h	49 f–h	28 gh	31.5 E
Mean	0.0 R	82.4 Q	100.3 P	91.8 PQ	

[a]Gx12, Gx26, Gx31 and Gx43 were *G. xylarioides* isolates obtained from Bebeka, Teppi, Jimma and Gera, respectively.
[b]Cultivars 74165, F-17 and 35/85 were resistant (R); 7440, F-61 and 61/85 were moderately resistant (MR); 74304, SN-5 and 24/85 were susceptible (S) to CWD under field conditions.
[c]0.0 indicates no external symptoms were observed over the test period. Means followed with the same letter(s) are not significantly ($P < 0.05$) different from each other according to Duncan's multiple range test. Least significant difference values for the cultivars, the isolates and the interactions comparisons were 13.7, 17.5 and 55.6, respectively.

74165 (40.5%) than Gx43 (Gera isolate), suggesting that the Teppi isolate was more aggressive than the Gera isolate (Gx43) on these cultivars, although Gx26 was less aggressive than the Gera isolate (Gx43) on cultivar F-61. The Jimma isolate (Gx31) was moderately aggressive on cultivars 74165, 74304, F-17 and SN-5, whereas the Bebeka isolate (Gx12) was least aggressive in all the cultivars (Table 8.3) (Girma and Mengistu, 2000; Girma and Hindorf, 2001).

The results concur with the conclusions of Zadoks and Van Leur (1983) but contradicted the findings of Pieters and Van der Graaff (1980). These latter authors determined percentage germination of conidia from four *G. xylarioides* isolates on the wood of branch internodes of six *C. arabica* varieties and concluded the presence of horizontal resistance (non-significant interaction). However, Zadoks and Van Leur (1983) found significant cultivar–isolate interaction indicating vertical resistance after re-analysing the same data reported by Pieters and Van der Graaff (1980). Thus, our results corroborated the existence of variation in both levels of resistance of coffee genotypes and in the aggressiveness of pathogen strains. A significant cultivar–isolate interaction (differential effect) further evidenced the existence of some vertical resistance in the *C. arabica* and *G. xylarioides* pathosystem (Girma and Mengistu, 2000; Girma *et al.*, 2001). Further investigations were then undertaken to compare fungal isolates from *C. arabica* and *C. canephora* to examine any host specificity in the host–pathogen interactions.

8.4. Interactions Among *C. arabica* and *C. canephora* vs. *G. xylarioides* Isolates From Both Host Species

In this study, ten isolates from *C. arabica* trees in Ethiopia were compared with an isolate from *C. canephora* in Uganda. The *C. arabica* isolates represented ten major arabica coffee-growing districts in Ethiopia, at varying altitudes (1000 to 2000 m above sea levels) and production systems, *vis.* semiforest, garden and plantation coffee. The isolates were inoculated on coffee seedlings in two sets of experiments following the standard inoculation protocols. The data analyses showed highly significant ($P < 0.001$) differences among cultivars, isolates and cultivar–isolate interactions both in percentage seedling wilt and incubation period (even excluding *C. canephora* line and isolate) (Girma, 2004; Girma *et al.*, 2005).

C. canephora appeared to be very susceptible and was severely attacked (84.1%) by isolate GxU12 in a short incubation period of 44 days (Table 8.5). However, no deaths were observed when *C. canephora* seedlings were inoculated with any arabica isolates (Gx1–Gx9 and Gx11) – even up to 12 months

Table 8.5. Percentage seedling death[a] among *C. arabica* cultivars and one *C. canephora* line inoculated with 11 *G. xylarioides* isolates collected from various geographical origins.

	C. arabica cultivars[c]						
Isolates[b]	Catimor-J19	7440	F-59	Caturra Rojo	24/85	*C. canephora*	Mean
Gx1	30.6 p–r	66.2 f–j	90.0 a	83.5 a–c	78.2 a–f	0.0 v	58.1 B
Gx2	19.9 r–u	52.5 j–n	78.2 a–f	81.7 a–d	69.6 c–i	0.0 v	50.3 C
Gx3	17.4 r–u	30.8 p–r	64.6 f–j	64.9 f–j	50.3 k–o	0.0 v	38.0 E
Gx4	27.9 q–s	65.8 f–j	83.9 ab	85.7 ab	80.3 a–e	0.0 v	57.3 B
Gx5	8.8 uv	30.5 p–r	67.3 e–i	47.6 l–o	44.6 m–o	0.0 v	33.2 E
Gx6	8.3 uv	42.3 n–p	77.4 a–f	65.2 f–j	68.4 d–i	0.0 v	43.6 D
Gx7	24.3 r–t	62.7 g–k	81.7 a–d	81.8 a–d	68.9 d–i	0.0 v	53.2 BC
Gx8	14.4 tu	27.1 q–t	75.0 b–g	58.7 h–l	38.0 o–q	0.0 v	35.5 E
Gx9	15.0 s–u	57.5 i–m	85.7 ab	81.6 a–d	62.5 g–k	0.0 v	50.4 C
Gx11	77.2 a–f	86.0 ab	81.5 a–d	90.0 a	72.2 b–h	0.0 v	67.8 A
GxU12	0.0 v	0.0 v	0.0 v	0.0 v	0.0 v	84.1 ab	14.0 F
Mean	22.2 T	47.4 S	71.4 P	67.4 Q	57.6 R	7.6 U	

[a]Percentage death was calculated from cumulative number of dead over total number of seedlings (20 per treatment) 6 months after inoculation, and the actual wilt values were arcsine-square root transformed to normalize the data. Means followed with the same letter(s) are not significantly different from each other. Least significant difference values ($P = 0.05$) for the cultivars, the isolates and the interactions comparisons are 3.5, 4.7, and 11.6, respectively. Coefficient of variation (CV) = 15.8%.

[b]Gx1, Gx2, Gx3, Gx4, Gx5, Gx6, Gx7, Gx8, Gx9, Gx11 and GxU12 designate *G. xylarioides* isolates collected from Jimma, Gera, Chira, Gechi, Yayu, Mettu, Tepi, Bebeka, Ayraguliso, Yirgacheffe and Uganda (*C. canephora* strain), respectively.

[c]Cultivars Catimor-J19 and 7440 were resistant; F-59 was moderately resistant; and Caturra Rojo and 24/85 were susceptible to CWD under field and greenhouse conditions. *C. canephora* was not affected by the disease in the field in Ethiopia.

after the trial had been completed. In contrast, *C. arabica* isolates were compatible with seedlings of all the arabica cultivars and caused varying percentages of death (Table 8.5) but induced no symptoms on *C. canephora* seedlings (Plate 17). The compatible–incompatible interactions of the isolates with the respective *Coffea* spp. were confirmed by re-isolating the fungus from the inoculated coffee seedlings. Results of both sets of the experiment proved host specificity or specialization of *G. xylarioides* populations to each *Coffea* spp. (Girma, 2004; Girma *et al.*, 2005). Among the *C. arabica* cultivars, Catimor-J19 showed a significantly ($P < 0.05$) lower mean percentage of dead seedlings (22.2%) followed by cultivar 7440 (47.4%) (Table 8.5), with incubation periods of 70 and 83 days, respectively, indicating high and moderate levels of resistance. Cultivars F-59 and Caturra Rojo were susceptible to the disease with the highest seedling deaths of 71.4% and 67.4%, respectively (Table 8.5) (Girma, 2004; Girma *et al.*, 2005).

There were significant ($P < 0.05$) ranges of variation in aggressiveness among the isolates of *C. arabica*. Isolates Gx3, Gx5 and Gx8 caused low seedling infections of 38.0%, 33.2% and 35.6%, respectively, with long incubation periods (Table 8.5). On the contrary, significantly high mean death rates of 58.1%, 57.3% and 67.8%% were induced by isolates Gx1, Gx4 and Gx11 (Table 8.5), respectively, suggesting that the former group of isolates were more aggressive than the latter group (Gx3, Gx5 and Gx8). The differential effects (interactions) indicated relatively low levels of infections on cultivar 7440 with Gx3, Gx5 and Gx8, with respective wilt incidences of 30.8%, 30.5% and 27.1%; whereas death rates of 66.2%, 65.8%, 62.7% and 86.0% were caused on the same cultivar with isolates Gx1, Gx4, Gx7 and Gx11, respectively (Table 8.5). Catimor-J19 showed some horizontal resistance, although it was highly infected by one isolate (Gx11). Cultivar 7440 was moderately resistant, whereas Caturra Rojo and 2485 were susceptible to most arabica isolates, but were moderately tolerant to Gx3, Gx5 and Gx8 isolates (Table 8.5) (Girma, 2004; Girma *et al.*, 2005).

In conclusion, in both sets of experiments, all *G. xylarioides* isolates derived from *C. arabica* induced disease in seedlings of cultivars of *C. arabica* (with varying levels of aggressiveness), but these isolates did not induce disease in *C. canephora*. Conversely, the strain from *C. canephora* was highly pathogenic to seedlings of its host but not to the *C. arabica* cultivars. Re-isolation from samples of inoculated seedlings of *C. arabica* and *C. canephora* with the respective strains showed a similar trend. These experiments represent the first time that cross-inoculation experiments between *C. arabica* and *C. canephora* plant material with their respective fungal isolates were undertaken and clearly evidenced host specialization of *G. xylarioides* (*F. xylarioides*) populations to the two commercial coffee species. Such host specificity has been previously speculated from field observations in DR Congo and Uganda (Flood, 1996, 1997) and in Ethiopia (Girma, 1997; Girma *et al.*, 2001). In addition to the indication of host specificity, the results further support the previous findings of Girma and Mengistu (2000) confirming horizontal resistance in *C. arabica*, aggressiveness in arabica isolates and vertical resistance/virulence in the host–pathogen combinations. However, because only one isolate of *G. xylarioides* from *C. canephora* had been studied, further investigations were conducted with more strains.

8.5. Interactions Among *Coffea* Species (*C. arabica/C. canephora/ Coffea excelsa*) With *G. xylarioides* Isolates From Either Host Species

The objectives of this study were to determine regional diversity of the current and historical strains of the coffee wilt pathogen collected from *Coffea* spp. in almost all of the regions where tracheomycosis has been a serious problem (Tables 8.6–8.8) and to further verify the reactions of the respective host species and host specialization of the fungus populations. Three independent sets of cross-inoculation experiments were conducted using recently isolated strains from *C. arabica* in Ethiopia (G3P22) and from *C. canephora* in Uganda (CAB003), DR Congo (RDC002) and Tanzania (TZ008, TZ009). A historical strain (DSMZ 62457) collected from *C. excelsa* in the Central African Republic (CAR) in the 1960s and a recent isolate (OUG152) from the same host species in Uganda were included. Seedlings of six *C. arabica* genotypes obtained from Ethiopia, Kenya and Costa Rica; eight *C. canephora* lines from DR Congo and Ivory Coast and *Coffea liberica* in Costa Rica were inoculated with suspensions of each isolate following the standard stem nicking method and then kept in a growth room with 12-h light/dark cycle at 25°C temperature throughout the study period (Girma *et al.*, 2007).

Table 8.6. Percentage seedling death[a] of *C. arabica* and *C. canephora* inoculated with *G. xylarioides* strains.

		G. xylarioides strain[c]			
	Cultivars/	Arabica strain	Canephora strain		Excelsa strain
Coffea spp.	Lines[b]	G3P22 (E)	CAB003 (U)	TZ009 (T)	DSMZ62457 (C)
C. arabica	K7	90.0	0.0	0.0	58.3
	SL28	93.8	0.0	0.0	62.5
C. canephora	LR/R1P2 (7)	0.0	61.3	34.5	13.3
	LR/R1P3 (17)	0.0	52.9	80.6	6.9
	LR/R1P4 (25)	5.4[d]	73.3	61.9	5.6
	TR CI17/37	0.0	20.6	14.3	0.0

[a]Percentage wilt was calculated from cumulative number of dead over total number of seedlings (ten per treatment) 6 months after inoculation.
[b]*C. arabica* cultivars (K7 and SL28) widely grown in Kenya and *C. canephora* lines obtained from DR Congo and Ivory Coast.
[c]*G. xylarioides* strains collected from *C. arabica* (arabica strain) in Ethiopia (E), *C. canephora* (canephora strains) in Uganda (U) and Tanzania (T) and a historical strain from *C. excelsa* (excelsa strain) in CAR (C).
[d]*G. xylarioides* was not re-isolated from seedlings (seedling death by other factors).

Table 8.7. Percentage seedling death[a] of *C. arabica* and *C. canephora* inoculated with different *G. xylarioides* strains.

		G. xylarioides strains[c]				
		Arabica strain	Canephora strain			Excelsa strain
Coffea spp.	Cultivars/ Lines[b]	G3P22 (E)	CAB003 (U)	TZ008 (T)	RDC002 (C)	OUG152 (U)
C. arabica	7454	53.3	0.0	0.0	0.0	0.0
	74165	30.0	0.0	0.0	0.0	0.0
C. canephora	TR CII7/1	4.0[d]	63.0	48.1	59.3	65.4
	TR CI17/16	8.3[d]	21.7	38.5	46.2	40.9

[a]Percentage wilt was calculated from cumulative number of dead over total number of seedlings (ten per treatment) 6 months after inoculation.
[b]*C. arabica* cultivars (7454 and 74165) grown in Ethiopia and *C. canephora* lines obtained from DR Congo.
[c]*G. xylarioides* strains collected from *C. arabica* (arabica strain) in Ethiopia (E); *C. canephora* (canephora strains) in Uganda (U), Tanzania (T), and DR Congo (C) and a recent strain from *C. excelsa* (excelsa strain) in Uganda (U).
[d]*G. xylarioides* was not re-isolated from seedlings (seedling death by other factors).

In the first trial set, *C. arabica* isolate G3P22 (IMI 392680) from Ethiopia was shown to be aggressively pathogenic to seedlings of the two *C. arabica* cultivars K7 and SL28 from Kenya, with 90% and 94% seedling deaths, respectively, at the end of the trial after 6 months (Table 8.6). Historical strain DSMZ 62457 (IMI 127629) of *C. excelsa* in CAR also infected seedlings of these cultivars (58% and 63% deaths, respectively). *C. canephora* isolates CAB003 (IMI 392263) and TZ009 (IMI 392679), obtained from recent CWD outbreak areas in Uganda and Tanzania, respectively, attacked the four *C. canephora* (robusta lines) from DR Congo and Ivory Coast with seedling mortality ranging from 14.3% to 80.6% (Table 8.6). Neither of these isolates induced symptoms in *C. arabica* (Plate 18). A number of the *C. canephora* seedlings inoculated with the historical strain also developed symptoms (13.3% death) (Girma *et al.*, 2007). Similar host–pathogen combinations were observed on grown seedlings (9 months old) of the catimor line (1570) inoculated with the three strains DSMZ, G3P22 and CAB003 (Plate 19).

In the second trial set, *C. arabica* isolate G3P22 infected seedlings of the two arabica coffee cultivars 7454 and 74165, which were obtained from Ethiopia, but did not induce symptoms on *C. canephora* lines (Table 8.7). Conversely, these *C. arabica* cultivars were not infected by any pathogen strains from *C. canephora*, namely, CAB003, TZ008 (IMI 392679) and RDC002 (IMI 392268). Strain OUG152 (IMI 392681), isolated from symptomatic excelsa coffee at Kituza, Uganda, caused up to 65% seedling death in *C. canephora* (robusta lines) but did not cause wilting in *C. arabica* (Girma *et al.*, 2007). In the third set of experiments, the arabica isolate G3P22 killed seedlings of the two *C. arabica* accessions known as Yemen/Java and E-238, which were

Table 8.8. Percentage seedling death[a] of *C. arabica*, *C. canephora* and *C. liberica* inoculated with *G. xylarioides* strains.

		G. xylarioides strains[c]		
Coffea spp.	Cultivars/ Lines[b]	Arabica strain G3P22 (E)	Canephora stain CAB003 (U)	Excelsa strain DSMZ62457 (C)
C. arabica	Yemen/Java	90.9	0.0	4.2[d]
	E-238	77.8	0.0	37.0
C. canephora	KR 10/7	0.0	51.9	15.4
(robusta type)	KR 15/4	0.0	42.3	0.0
C. liberica	T-1984	5.0[d]	100	80.0
	T-1872	21.1	95.8	94.4

[a]Percentage wilt was calculated from cumulative number of dead over total number of seedlings (ten per treatment) 6 months after inoculation.
[b]*C. arabica* cultivars (Yemen/Java and E-238) grown in Costa Rica, *C. canephora* (robusta line) obtained from DR Congo and *C. liberica* collected in Costa Rica.
[c]*G. xylarioides* strains collected from *C. arabica* (arabica strain) in Ethiopia (E), *C. canephora* (canephora strains) in Uganda (U) and a historical strain from *C. excelsa* (excelsa strain) in CAR (C).
[d]*G. xylarioides* was not re-isolated from seedlings (seedling death by other factors).

received from Costa Rica, with 91% and 78% death, respectively (Table 8.8). Seedlings of *C. liberica* lines, T-1984 and T-1872, from the same country were susceptible to the historical strain DSMZ 62457 and the current robusta isolate CAB003. The characteristic wilting symptoms appeared on these strain–coffee line combinations within less than 30 days of inoculation, and most of the seedlings collapsed at cotyledon stage (Plate 20). Accession T-1872 also exhibited moderate susceptibility to the arabica isolate.

In summary, there were no wilting symptoms, and the fungus was also not re-isolated from control (uninoculated) seedlings in the three trial sets; successful re-isolation of *G. xylarioides* from plant tissues of most of the inoculated and symptomatic seedlings (inoculated with their respective isolates) confirmed the host–pathogen compatibility. The results of all these cross-inoculation trials further evidenced that, with the exception of the historical strain, host specialization occurs in the *G. xylarioides* populations. Those strains collected in the field from *C. arabica* trees showing CWD symptoms were pathogenic to *C. arabica* cultivars obtained in Ethiopia, Costa Rica and Kenya and suggest that this disease could be a serious problem on arabica coffee if introduced into countries that are tracheomycosis-free (both in Africa and beyond). In addition, the strains isolated from *C. canephora* trees in countries with recent CWD outbreak, namely, DR Congo, Uganda and Tanzania, were aggressively pathogenic to all the seedlings of *C. canephora* (robusta lines) collected in Ivory Coast and DR Congo (Girma *et al.*, 2007).

The recent strain OUG152 (IMI 392681) isolated from *C. excelsa* in a clonal trial site of the Coffee Research Institute at Kituza, Uganda, also induced disease in *C. canephora* lines but did not affect *C. arabica*. Excelsa strains

OUG151 (IMI 392265), OUG154 (IMI 392682) and OUG155 (IMI 392266) isolated from the same host species showed varying levels of seedling death on *C. canephora* (robusta line, 1331) (D. Bieysse, personal communication). In contrast, the historical excelsa strain DSMZ 62457 (IMI 127629), originally collected from *C. excelsa* trees in CAR in the 1960s, induced typical CWD symptoms and exhibited a wide range of pathogenicity (seedling death) on *C. arabica*, *C. canephora* and *C. liberica*. Pathogenicity tests of other historical strains ATTC 15664 (IMI 392676) and CBS 749.79 (IMI 392675) collected from *C. canephora* in Ivory Coast and CBS 258.52 (IMI 392674) isolated from an unknown *Coffea* sp. in Guinea caused death of 10%, 30% and 35% of plants of *C. canephora* (robusta line 1406), respectively (D. Bieysse, personal communication). These results suggest that historical strains collected during the 1960s remained aggressively pathogenic to *Coffea* spp. after 40 to 50 years of preservation. Girma *et al.* (2005) hypothesized that the compatibility and infection of *C. canephora* by excelsa strains imply that *G. xylarioides* population presently causing CWD outbreaks in DR Congo, Uganda and Tanzania may have arisen from older populations in CAR or, alternatively, that a separate divergent population is evolving on *C. excelsa* in nature (Girma *et al.*, 2007).

8.6. Histological Studies of *C. canephora*/*G. xylarioides* Interactions

Detailed understanding of infection processes from penetration of the pathogen to colonization of the host tissues is very important to define host–pathogen interaction at cellular levels. Thus, preliminary observations were carried out on *C. canephora* (robusta line) seedlings 9 and 18 months old that were inoculated with *G. xylarioides* strain CAB003 (IMI 392263) by injection of a droplet of spore suspension of conidia (10^6 conidia/ml) into the trunk under the cotyledon leaves using a syringe. Light microscopic studies were carried out on 2-µm sections excised from the inoculated seedlings showing early symptoms (leaves were partially wilting) 8 weeks after inoculation. The five internodes were sampled successively upwards starting from the cotyledon leaves per sample seedlings.

8.6.1. Histological study on 9-months-old *C. canephora* seedlings

Microscopic observations of inoculated plants at the first internode level above the cotyledons revealed that the epidermis and external parenchyma had disappeared and the cork and phellogen layers had partially disintegrated. The wood tissues were found disturbed in some places but had not become totally disorganized; no mycelium could be seen in these sections. The epidermis, external parenchyma and phellogen were all absent at the 2nd internode position, and slight hypertrophy of the parenchyma was observed, but the xylem tissue was intact and no mycelium was detected. At the 3rd internode level, the epidermis was absent, and swelling of the external and upper parenchyma cells had produced a clear tissue deformation and some disintegration. Xylem and pith cells had also disintegrated and contained a large quantity of starch grains, but no mycelium was observed. At the 4th internode, a swollen discoloured structure was observed, which spread over 1 cm along the stem. The swelling of the epidermis and external parenchyma cells was due to an increase of the volume of the vacuole. Although phellogen cells were damaged, the cork cell structure remained intact except for a slight swelling of cell nuclei. The fungus was observed in the wood tissues, more especially in fibres, and appeared as a blue, intracellular mycelium colonizing fibres closer to the pith (linearly following the rays) (Plate 21). Cells that contained mycelium had little or no lignin on their walls. Mycelial colonization appeared absent from vessels. At the 5th internode, the youngest level, the epidermis was absent; parenchyma cells were strongly hypertrophied leading to tissue distortion; the phellogen was absent and no mycelium observed. A significant accumulation of starch granules could be observed in pith and in the wood cells. The presence of mycelium at the 4th node from the infection point indicates that it had colonized the plant probably through translocation of conidia that are transported in the sap towards the apical part of the plant.

8.6.2. Histological interaction study on 18-month-old *C. canephora* seedlings

In 18-month-old plants, where the symptoms were advanced, the mycelium was observed to be invading the xylem, and dense colonization was seen in the vessels and in the intercellular spaces (Plate 22). However, in sections taken from those plants exhibiting early symptoms, the fungus was not observed.

8.7. Summary and Conclusions

The prospects of successful control of CWD rely principally upon deployments of resistant coffee cultivars/lines. In this regard, full-fledged, independent, large-scale coffee collection and screening strategy should be planned hand in hand with efficient breeding programme. Because of the difficulty in screening for resistance under field conditions for soil-borne diseases, standard seedling inoculation protocols should be employed. For a number of technical reasons, stem nicking (at 2 cm above the soil level) of late-cotyledon-stage (2 to 2.5 months old) coffee seedlings with inoculum concentration of 2×10^6 conidia/ml is recommended. The inoculated seedlings need to be placed in an air-conditioned room with higher humidity (>95%) and temperature of about 23°C for 1 week that ensure infection and then be transferred to the greenhouse (Plate 23). Disease parameters such as number of dead seedlings and number of days between inoculation and external symptom appearance (chlorosis, retarded plant growth and reduced leaf size, wilting and finally death) should be recorded for at least 6 months (preferably at fortnightly intervals), from which percentage death and incubation periods can be computed and used for statistical analyses. Pieters and Van der Graaff (1980) reported the highest correlation between the field scores and death rates and incubation time in the seedling tests 6 months after inoculation. The resistance of surviving seedlings of a particular cultivar/line needs to be further verified by re-inoculating grown seedlings (12 months old) in the screen house (following the same technique) and evaluated at least for one more year (Plate 24) because of the difficulties of proving host resistance under field conditions.

Following on from the detailed host–pathogen interactions supported by random amplified polymorphic DNA analysis, Girma *et al.* (2005) introduced the epithet *formae speciales* (special forms), *G. xylarioides* f. sp. *abyssiniae* (anamorph: *F. xylarioides* f. sp. *abyssiniae*) for strains attacking *C. arabica* confined to Ethiopia and *G. xylarioides* f. sp. *canephorae* (anamorph: *F. xylarioides* f. sp. *canephorae*) for strains specifically pathogenic to *C. canephora* and *C. excelsa*. Rutherford (2006), based on the various molecular studies, reported that two clonal populations are responsible for the current CWD outbreaks in Africa, one composed of isolates obtained from affected *C. arabica* in Ethiopia (variant 'A'), and the other are isolates from affected *C. canephora* in DR Congo, Uganda, and Tanzania (variant 'C') (Chapter 7). Phylogenetic analysis of *F. xylarioides* including the number of isolates obtained from CWD-affected coffee trees in Uganda has shown these to belong to the African clade of the *Gibberella fujikuroi* species complex, a clade replacing *Fusarium* section *Liseola* (Geiser *et al.*, 2005). The appearance of new populations or strains of the pathogen also needs to be monitored in the field because *G. xylarioides* produces very abundant perithecia and ascospores in the stem bark of dead trees and stumps of coffee (Van der Graaff and Pieters, 1978; Flood and Brayford, 1997; Girma *et al.*, 2001; Girma, 2004).

The preliminary histopathological observation of the interaction of *C. canephora* vs. *G. xylarioides* highlighted the hyphal colonization of xylem tissues and the intercellular spaces, blocking water movement that leads finally to wilting of infected coffee trees. However, the underlying resistance mechanisms that may be morphological, physiological and/or biochemical involved in host defense systems of different cultivars/lines of *Coffea* species should also thoroughly be further studied. The resistance in *C. canephora* and *C. liberica* was supposed to be associated with rapid suberization in wounds and occurrence of caffeine and chlorogenic acid in high concentrations (Booth, 1971; Holliday, 1980). Understanding the gene(s) that govern resistance to CWD and mode of inheritance of the resistance mechanisms is of paramount importance to design and implement successful breeding programme.

Acknowledgements

The authors acknowledge the Department for International Development, UK, for funding part of the work done at CIRAD, Montpellier, and we are very thankful to all the staff at Jimma Agricultural Research Center (EIAR, Ethiopia) and CIRAD (France) for their unreserved technical assistance.

References

Beckman, C.H. (1987) *The nature of wilt diseases of plants.* APS Press, St. Paul, MN, USA.

Beckman, C.H. and Talboys, P. W. (1981) Anatomy of resistance. In: Mace, M.E., Bell, A.A. and Beckman C.H. (eds.) *Fungal Wilt Diseases of Plants.* Academic Press, New York, pp. 487–521.

Booth, C. (1971) *The Genus Fusarium.* Commonwealth Mycological Institute, Kew, Surrey, England.

Flood, J. (1996) A study of tracheomycosis or vascular wilt disease of coffee in Zaire. Report presented to Zairean Coffee Organization (OZACAF) August 1996, 13 pp.

Flood, J. (1997) Tracheomycosis or vascular wilt disease of coffee in Uganda. Report presented to Ugandan Coffee Development Authority (UCDA), Uganda. 12 pp.

Flood, J. (2006) A review of *Fusarium* wilt of oil palm caused by *Fusarium oxysporum* f. sp. *elaeidis. Phytopathology* 96, 660–662.

Flood, J. and Brayford, D. (1997) Reemergence of *Fusarium* wilt of coffee in Africa. In: *Proceedings of the 17th International Scientific Conference on Coffee Science (ASIC).* Nairobi, Kenya, pp. 621–627.

Geiser, D.M., Lewis Ivey, M.L., Hakiza, G., Juba, J.H. and Miller, S.A. (2005) *Gibberella xylarioides* (anamorph: *Fusarium xylarioides*), a causative agent of coffee wilt disease in Africa, is a previously unrecognized member of the *G. fujikuroi* species complex. *Mycologia* 97, 191–201.

Girma, A. (1997) Status and economic importance of *Fusarium* with disease of Arabica coffee in Ethiopia. In: Hakiza, G. J., Birkunzira, B. and Musoli, P. (eds.) *Proceedings of the First Regional Workshop on Coffee Wilt Disease (Tracheomycosis).* International Conference Centre, Kampala, Uganda, pp. 53–61.

Girma, A. (2004) Diversity in pathogenicity and genetics of *Gibberella xylarioides* (*Fusarium xylarioides*) populations and resistance of *Coffea* spp. in Ethiopia. PhD dissertation. University of Bonn, Bonn, Germany.

Girma, A., Flood, J., Hindorf, H., Bieysse, D., Simons, S. and Mike, R. (2007) Tracheomycosis (*Gibberella xylarioides*) – a menace to world coffee production: evidenced by cross inoculation of historical and current strains of the pathogen. In: *Proceedings of the 21st International Scientific Conference on Coffee Science (ASIC).* Montpellier, France, pp. 1268–1276.

Girma, A. and Hindorf, H. (2001) Recent investigation on coffee tracheomycosis, *Gibberella xylarioides* (*Fusarium xylarioides*) in Ethiopia. In: *Proceedings of the 19th International Scientific Conference on Coffee Science (ASIC).* Trieste, Italy, pp. 1246–1252.

Girma, A., Hindorf, H., Steiner, U., Nirenberg, H., Dehne, H.-W. and Schellander, K. (2005) Genetic diversity in the coffee wilt pathogen (*Gibberella xylarioides*) populations: differentiation by host specialization and RAPD analysis. *Journal of Plant Diseases and Protection* 112, 134–145.

Girma, A. and Mengistu, H. (2000) Cultural characteristics and pathogenicity of *Gibberella xylarioides* isolates on coffee. *Pest Management Journal of Ethiopia* 4, 11–18.

Girma, A., Mengistu, H. and Hindorf, H. (2001) Incidence of tracheomycosis, *Gibberella xylarioides* (*Fusarium xylarioides*) on arabica coffee in Ethiopia. *Journal of Plant Diseases and Protection* 108, 136–142.

Hakiza, G.J., Kyetere, D.T. and Olal, S. (2004) Mode of penetration and symptom expression in robusta coffee seedlings, inoculated with *Gibberella xylarioides*, the cause of coffee wilt disease in Uganda. In: *Proceedings of the 20th International Scientific Conference on Coffee Science (ASIC)*. Bangalore, India, pp. 1232–1233.

Holliday, P. (1980) *Fungal Diseases of Tropical Crops*. Cambridge University Press, London.

Kilambo, D.L., Ng'homa, N.M., Mtenga, D.J., Teri, J.M., Nzallawahe, T., Mike, R. and Masumbuko1, L. (2007) Progress towards searching for durable resistance to *Fusarium* wilt (*Fusarium xylarioides*) in *Coffea canephora* germplasm in Tanzania. In: *Proceedings of the 20th International Scientific Conference on Coffee Science (ASIC)*. Bangalore, India, pp. 1386–1389.

Kranz, J. (1962) Coffee diseases in Guinea. *FAO Plant Protection Bulletin* 10, 107–109.

Leung, H., Nelson, R.J. and Leach, J.E. (1993) Population structure of plant pathogenic fungi and bacteria. *Advances in Plant Pathology* 10, 157–205.

McDonald, B.A. (1997) The population genetics of fungi: tools and techniques. *Phytopathology* 87, 448–453.

Muller, R.A. (1997) Some aspects of past studies conducted in western and central Francophone Africa on coffee tracheomycosis. In: Hakiza, G. J., Birkunzira, B. and Musoli, P. (eds.) *Proceedings of the First Regional Workshop on Coffee Wilt Disease (Tracheomycosis)*. International Conference Centre, Kampala, Uganda, pp. 18–30.

Oduor, G., Simons, S., Phiri, N., Njuki, J., Poole, J., Pinard, F., Kyetere, D., Hakiza, G., Musoli, P., Lukwago, G. Abebe, M., Tesfaye, A., Kilambo, D., Asiimwe, T. and Munyankere, P. (2003) Surveys to assess the extent and impact of coffee wilt disease in East and Central Africa. Final Technical Report EU contract no. ASA-RSP/CV-006. CAB International, Egham, UK.

Pieters, R. and Van der Graaff, N.A. (1980) *Gibberella xylarioides* on arabica coffee: evaluation of testing methods and evidence for the horizontal nature of resistance. *Netherlands Journal of Plant Pathology* 86, 37–43.

Rutherford, M.A. (2006) Current knowledge of coffee wilt disease, a major constraint to coffee production in Africa. *Phytopathology* 96, 663–666.

Van der Graaff, N.A. and Pieters, R. (1978) Resistance levels in *Coffea arabica* L. to *Gibberella xylarioides* and distribution pattern of the disease. *Netherlands Journal of Plant Pathology* 84, 117–120.

Van der Plank, J.E. (1984) *Disease Resistance in Plants*, 2nd edn. Academic Press, London.

Zadoks, J.C. and van Leur, J.A. (1983) Durable resistance and host–pathogen–environment interaction. In: Lamberti, L., Waller, J.M. and Van der Graaff, N.A. (eds.) *Durable Resistance in Crops*. Plenum Press, New York, pp. 125–140.

Watson, I.A. (1970) Changes in virulence and populations shifts in plant pathogens. *Annual Review of Plant Pathology* 8, 209–230.

Wellman, F.L. (1961) *Coffee Botany, Cultivation and Utilization*. Leonard Hill (Books) Ltd., London.

9 Management of Coffee Wilt Disease

N. Phiri, M. Kimani, E. Negussie, S. Simons and G. Oduor
CABI Africa, UN Avenue, ICRAF Complex, PO Box 633-00621, Nairobi, Kenya

9.1. Introduction

Until the early 1990s, coffee (*Coffea* spp., *Rubiaceae*) was the world's most important internationally traded commodity in terms of monetary value, after oil (Rice and Ward, 1996). Currently, however, coffee is ranked as only the 5th most important traded commodity after oil, aluminium, wheat and coal. In 2003, global green coffee production was 7,221,255 t, of which approximately 60% was arabica (*Coffea arabica*) and 39% robusta coffee (*Coffea canephora*) (Rice and Ward, 1996). Of the coffee produced in 2003, 5,233,064 t were exported, with a value of approximately $5.5 billion (FAO, 2004). Despite its decline in rank as a traded commodity, coffee continues to be an important source of foreign exchange earnings, and it is the primary export of many developing countries. Cultivation, processing, trading, transportation and marketing of coffee provide employment for millions of people worldwide (International Coffee Organisation [ICO], 2005).

In sub-Saharan Africa, coffee is the economic backbone of more than 50 countries and central to the livelihoods of more than 20 million rural families (Oduor and Simons, 2003). It is also a major source of income for millions of smallholder coffee growers and their households who are responsible for an estimated 80% of coffee production in Africa (Oduor and Simons, 2003). Coffee is the most important cash crop for Africa as a whole, contributing some 10% of the total foreign exchange earnings in the continent (FAO, 2004). A number of coffee-producing countries in sub-Saharan Africa, including Uganda, Ethiopia, Rwanda and Burundi, depend on the export of this commodity for more than half of their foreign exchange earnings.

However, African coffee suffers from a range of co-evolved pests and diseases including coffee berry disease, coffee leaf rust (CLR), coffee berry borer etc. Global coffee production has nearly doubled since 1970, but Africa's share of total production declined from 30% to less than 15% (FAO,

2004) despite accounting for one third of the world's coffee hectarage. The average yields in Africa are generally low and declining, ranging from 0.3 to 0.38 t/ha, half that achieved in Latin America and almost one third of Asia's productivity. This is partly due to history: some 40% of African plantations date from the pre-independence era and have not been renewed since. The continued reliance on outdated and often unproductive varieties in the face of widespread prevalence of pests and diseases has contributed to this decline, as has the recent re-emergence of coffee wilt disease (CWD).

CWD or tracheomycosis is a highly destructive disease of robusta coffee (*C. canephora*) and arabica coffee (*C. arabica*) throughout East and Central Africa. Coffee production in Uganda and the Democratic Republic of Congo (DRC) has been decimated following the re-emergence of CWD in the 1970s and 1980s in DRC and 1990s in Uganda (Flood and Brayford, 1997; G. Hakiza, 1995, unpublished data; Oduor *et al.* 2003; Office Zairois du Cafe [OZACAF], 1995), and the disease has since spread to neighbouring countries in the region including Tanzania. CWD is caused by a fungus called *Gibberella xylarioides* R. Heim & Saccas, which has *Fusarium xylarioides* as its anamorphic phase. Following the re-emergence of CWD, income from coffee declined by over 50% in three CWD-affected countries. A number of control methods are available for the management of CWD, which include use of resistant varieties, chemical control and cultural methods.

This chapter discusses available management methods and those which could be available in the near future.

9.2. Cultural Practices Used in Managing CWD

A range of cultural practices can be used in managing CWD. Perhaps the commonest of all and a well-known method is the uprooting of infected coffee trees and burning *in situ*. Other cultural methods include quarantine (intra- and interfarm, inter- and intracountry quarantine), use of disease-free planting materials, prevention of coffee tree wounding, control of insect pests, disinfecting farm implements (e.g. secateurs, pruning saws) and general plant health. These methods are discussed below.

9.2.1. Uprooting and burning

Uprooting and burning is probably one of the oldest CWD management methods. It dates back to the first CWD outbreak in the 1950s (Saccas, 1956) and contributed to the control of CWD when coupled with use of resistant varieties. The method works by removing the source of pathogen inoculum and involves frequently inspecting the coffee farm to identify infected coffee bushes. Once identified, the infected coffee bushes are uprooted by digging out as much of the root system as possible then burning them on the spot at the earliest opportunity – in the hole where the coffee bush was uprooted (Plate 28). Although early identification of symptoms, uprooting and burning the infected coffee trees at this early stage is crucial for CWD control, it poses problems because the coffee materials are not dry enough at this stage and do not burn easily. The best way for carrying out uprooting and burning newly infected coffee trees is, therefore, to cut the coffee materials into smaller pieces, which promotes drying, and these pieces can then be heaped in the hole from where the infected coffee tree was uprooted. Dry firewood could also be used to facilitate the burning. To reduce further risk of inoculum in adjacent coffee trees, those coffee trees surrounding infected coffee trees also need to be uprooted and burnt. Uprooting the coffee tree that has already died from CWD is too late; the pathogen will have already spread to surrounding and distant coffee trees in the farm and even to adjacent farms. Conidia and ascospores of *G. xylarioides* spread through wind, rain and human activities (e.g. harvesting, pruning) (Jacques-Felix, 1954). When CWD infection levels are high, such as 70% of coffee trees being infected in a coffee garden, it is advisable to uproot and burn all coffee trees in the farm and replant with resistant or tolerant coffee germplasm. This may pose problems due to loss of the little remaining source of income for the farmer. Educating the farmers about the advantages and disadvantages of uprooting and burning all coffee trees in heavily infected farms should help in this case. The practice of burning infected coffee plant material on the spot helps in limiting the spread of *G. xylarioides* inoculum from the plant material and the contaminated soil. *G. xylarioides* is considered to be an endemic soil-inhabiting fungus (Chapter 7). Although most of the root system may be dug out, much still remains in the soil and is potential source of inoculum. Such inoculum can be eliminated by observing a fallow period of at least 6 months (Hakiza, personal communication) to 2 years (Wrigley, 1988) before replanting with a susceptible coffee seedling. By this time, the viability of the soil inoculum will have declined (Wrigley, 1988).

Training of farmers in the identification of CWD symptoms, especially the early stages of symptom expression, is crucial for early diagnosis and therefore for better chances of eradication. CWD symptoms include leaf curling, chlorosis and wilting (Plate 3), with eventual leaf defoliation leading to dieback (Plates 3 and 4). The symptoms, in most cases and especially in multiple-stem coffee bushes, start on one side of the coffee stem where the infested vascular bundles are, which have been blocked by a combination of fungal colonization and host responses. These symptoms are confirmed by scraping the diseased stem with a knife. A blue–black stain is characteristic of an infected coffee stem (Plate 8). Training of farmers and extensionists is so important in the management of CWD that it formed a major component of the Regional Coffee Wilt Programme (RCWP), Chapter 11.

Yet despite clear guidance about burning *in situ*, there is a common practice by people within the coffee-growing areas to transport infected coffee stems to their homesteads for use as firewood, as staking material for climbing beans and even as fencing material around coffee gardens (Chapters 2, 3, 4 and 5). These practices help spread CWD further in the area. This poses a huge threat to the spread of CWD, and limiting these practices will improve management of CWD. Proper disposal of infected coffee material is crucial in the management of CWD because the bark, roots and stems all contain the pathogen in the form of spores, hyphae and fruiting bodies (CAB International [CABI], 2006).

There are a number of examples of use of uprooting and burning. When CWD first appeared in West Africa from 1927 to the 1950s, it was controlled in Cameroon by mobilizing soldiers to uproot and burn all infected coffee in the country (Deassus, 1954). In recent outbreaks, CWD is being controlled by practicing uprooting and burning in addition to other control methods. In Tanzania, for example, farmers are carrying out uprooting and burning campaigns of CWD-infected coffee trees and are being supported by their extensionists, members of Tanzania Coffee Research Institute (TACRI) and the local, district and regional leaders in the Bukoba, Misenyi, Karagwe and Muleba districts of the Kagera region (Chapter 5). Uprooting and burning is used in combination with use of copper stem paint and mulching. In addition, uprooting and burning of infected coffee trees is also being practiced in DRC, Ethiopia and Uganda (Chapters 3 and 4).

However, where governments need to bring in soldiers or others to uproot and burn farmers' CWD-infected coffee as part of a concerted action, appropriate legislation to back up such an action may be needed, although many countries already have such legislation in place.

9.2.2. Quarantine as a tool for managing the spread of CWD

Quarantine is an important cultural control method for CWD and is aimed at preventing the movement of infected plant material, infected soil and infected implements to 'clean' areas where it can be a source of new disease outbreaks. Quarantine can be used at farm, national and international levels. It is rare for most people to think of quarantine at farm level. However, it is crucial that a farmer observes quarantine to limit spread of CWD in or among his/her farm(s). The practice includes carrying out interfarm activities from the non-affected farm and finishing with the affected farm. In addition, the farmer should have separate implements for the affected and non-affected farms to reduce the risk of spreading CWD to the clean farm. When working within the affected farm, the farmer should start working in the clean section and finish with the affected section of his/her coffee farm, thus minimizing spread of CWD within the farm. This is because the pathogen can spread through contaminated farm implements, even on clothes and shoes/feet when contaminated soil sticks to the shoes/feet or farm implements. Spores of *G. xylarioides* can also stick to clothes if a farmer is in contact with infected coffee bushes, especially if they have been left in the garden for a long time. Quarantine can also be used between farmers; exchange of coffee plant material and farm implements should be discouraged to reduce the risk of spreading CWD. The practice of using diseased material as firewood (see 9.2.1) has had a huge impact on the spread of CWD, and steps to limit these practices are being implemented – mainly through raising awareness among farmers.

Quarantine also involves monitoring and restricting movement of plant material between affected and non-affected countries. For example, coffee-growing countries like Burundi, Rwanda and Kenya, which have neighbours with CWD outbreak, should be diligent in preventing cross-border movement of coffee material (germplasm, coffee in parchment or green coffee) to prevent spread of CWD from the infected neighbouring countries. This has already occurred between DRC and Uganda. CWD was observed on abandoned coffee farms in eastern DRC in 1986 and is believed to have moved or been moved to Uganda, as it was recorded in the Ugandan district of Bundibujyo in the early 1990s (Flood and Brayford, 1997). However, implementation of quarantine measures at intercountry levels may be difficult to enforce in Africa because of the many porous borders and unmonitored informal cross-border trade, which could have accounted for the spread of the current outbreak of CWD from DRC to Uganda and subsequently from Uganda to Tanzania.

Enforcement of intercountry quarantine is probably the only chance for Kenya's *C. arabica*, which faces threat from the arabica strain of *G. xylarioides* that is affecting *C. arabica* in Ethiopia. Studies by Girma *et al.* (2001) (Chapter 8) showed that most of the existing Kenyan coffee varieties, except for Ruiru 11, which was not included in the study, were susceptible to the arabica strain, which is referred to as variant 'A'. However, the desert between the two countries provides a natural barrier and helps in reducing the risk of the arabica strain from spreading to Kenya and the rest of the arabica-coffee growing countries south of Ethiopia.

9.2.3. Use of disease-free planting materials

CWD is spread through planting materials, among others. Although it is a subject of debate, transmission through seed is a potential way of spreading CWD. Wrigley (1988) quotes Clowes and Hill (1981), who claimed that a berry and branch blight in Zimbabwe was due to *G. xylarioides* and that this was spread in seed from berries that may contain the CWD pathogen. However, CWD has not been confirmed in Zimbabwe, and further observations showed that the disease in Zimbabwe was similar to that caused by *Fusarium* bark disease (*Fusarium stilboides*) and was probably confused with CWD. Studies by Girma and Hindorf (2001) showed that seeds did not transmit CWD in arabica coffee in Ethiopia. It is important to note, however, that *G. xylarioides* is a systemic pathogen affecting the vascular elements, and therefore, seed transmission may be possible even though it has not been confirmed so far. Seedlings may also transmit *G. xylarioides* through soil adhering to their roots and spores carried on their stems and branches. Seedlings from reliable sources, such as registered clonal propagators, may limit this risk. In addition, farmers and seedling producers should be trained in using potting soil from an area without the potential of harbouring *G. xylarioides*, such as a virgin land or an area without a history of coffee. Soil sterilization is another option for eliminating soil-borne inoculum, but this may be expensive and time consuming.

The availability of coffee planting materials is a challenge at the moment because of lack of available varieties or clones resistant to CWD. The current available resistant clones in Uganda are still being propagated, and it will be sometime before these clones will be widely available to farmers. Resistant clones in Tanzania and DRC are yet to be evaluated for other traits such as red blister (*Cercospora coffeicola*) and CLR and agronomic traits.

9.2.4. Prevention of coffee tree wounding

G. xylarioides penetrates through wounds, so any agency causing wounds will aid the spread of the fungus. Kranz and Mogk (1973) noted that most dying and dead trees had been wounded during weeding. The main author observed wounds caused by weeding practices, in particular slashing with a bushman's knife (slasher) on most of the infected coffee trees in Ethiopia (Plate 6). This is worsened by the fact that most of the coffee are grown using very close spacing in Ethiopia, and some of the coffee are actually semiforest type, i.e. it was not planted by farmers, hence may grow at even closer spacing than the recommended one. Coffee tree wounding is also prevalent in most countries in Africa, and it is mainly due to weed management.

In addition, there are other coffee farm operations that have to be carried out which cause wounds on the coffee bush. Particular operations that pose a huge challenge in the prevention of CWD spread is the rejuvenation of coffee or changing of a coffee cycle that involves stumping. Changing a coffee cycle is a routine practice that has to be carried out when coffee bushes have aged or have become unproductive. The operation requires cutting of coffee stems at about 45 cm above the soil level, at an angle, with a pruning saw so that new coffee shoots that sprout below the cut point are selected and one or two are allowed to form the next coffee cropping cycle. Changing the coffee cycle is carried out several times for over a 50-year period, and the technique allows the bushes to be maintained over many years. Although this is a very important way of turning unproductive coffee bushes into productive ones, there is a high risk of transmitting *G. xylarioides* through the pruning saw. In addition, the freshly cut surface of the coffee stump provides a very large surface area that can be used as an entry point for *G. xylarioides*. The main author here has observed farms where coffee trees were completely destroyed by CWD after changing the cycle in Ethiopia. The new suckers that were sprouting from the cut stumps were all infected with CWD and hence died before they even grew up to 1 m from the growing point. To prevent this catastrophe, the implements should be flamed over fire before cutting each stem or after cutting a few stems so that they are sterilized, particularly in farms with the history of CWD. Coffee trees showing CWD symptoms need to be uprooted and burnt. If a farmer can afford, it is advisable for the stumps to be painted with a fungicide paste. An ordinary copper-based fungicide can be used for this purpose. The copper-based fungicide should be mixed with water at the rate of 300 g copper-based fungicide to 1 l of water. The paste can be applied with a paint brush on the cut surface of the stump.

Pruning to cut off dead or interlocking branches or branches that are pointing inwards is also a potential way of spreading CWD and should be managed properly. Secatuers that are used in the pruning process should be sterilized over fire as described above.

However, wounding also occurs naturally at the collar area of the coffee stem (the area of the coffee stem at the soil level). These microscopic wounds develop mostly when the trees sway due to heavy winds. The wounds can therefore be reduced through provision of wind breaks, for example planting a band of bananas or closely planted trees, in particular fast-growing trees around the edge of the coffee farm, could help reduce the level of this type of wounds.

Another cause of wounding is wood-boring insect pests such as the white coffee stem borer (*Monochamus leuconotus* [Pascoe] or *Bixadus sierricola* [White]) and the yellow-headed borer (*Dirphya nigricornis* [Olivier]), which create entry points, ring barking and entry and exit holes that could provide entry points for *G. xylarioides*. In addition, insects are suspected to transmit CWD from tree to tree. Insect pests ought to be controlled to prevent the spread of CWD in addition to the primary objective of preventing direct crop losses caused by the insect pest damage. Fipronil, at a dilution rate of 1.25 ml in 1 l of water, was found effective in controlling the white stem borer in Malawi and Zimbabwe when applied as stem paint to the first 50 cm of the stem from the soil level. In addition, chlorpyrifos diluted at 35 ml in 1 l of water and applied to the first 50 cm of the coffee stem was also found effective to white stem borer in Zimbabwe (Integrated Stem Borer Management in Smallholder Coffee Farms in India, Malawi and Zimbabwe Project [CFC/ICO/18] Final Technical Report, 2008).

Livestock, in particular goats and cattle, also wound coffee stems. Goats were observed, on a number of occasions, eating the coffee bark in Uganda when left to graze in the coffee gardens. In addition, cattle were found tethered to coffee stems resulting in the rope, used for tethering the animals, wounding the coffee stem as the tethered animal moved around the coffee tree when grazing. It is therefore imperative that goats ought to be prevented from grazing in coffee gardens, and cattle should not be tethered to coffee stems to prevent wounding.

Wounding can be avoided by a number of careful management techniques that help in managing CWD.

9.2.5. Weed management practices for preventing coffee stem wounding

Mulch (Plate 30) suppresses weeds but limits the use of herbicides, and combined with hand-picking weeds around the coffee stems instead of slashing significantly reduced the incidence of CWD in recent CWD management studies, which were part of the Improvement of Coffee Production in Africa by the Control of Coffee Wilt Disease (Tracheomycosis) Project (CFC/ICO/13).

This resulted in farmers adopting these methods for controlling weeds. In addition, live mulch, *Desmodium intotum*, which suppresses weeds during its growing condition (Plate 31), was also adopted as an alternative to weed control after substantial studies by the weed scientist at Jimma Agriculture Research Centre in Ethiopia.

Trials on evaluation of cultural and chemical control methods for the management of CWD in Ethiopia, Uganda, Tanzania and DRC

Following biological and socio-economic surveys that were carried out in Uganda, Ethiopia, Tanzania and DRC, some factors from the surveys showed some positive correlation to CWD. It was therefore decided that the country-specific factors be evaluated on farm and on station in these countries. However, some management options also came from literature, scientists' personal experience and experience of farmers and extensionists. Management options that were agreed upon in participatory workshops attended by scientists, farmers and extensionists are presented in Table 9.1.

Details for applying the treatments:

- Herbicides (Roundup) – applied at 150 ml Roundup in 15 l of water with a sprayer. Applied when weeds appear (roughly applied twice a year).
- Copper oxychloride spray – mixed at the rate of 40 g per 15 l of water and sprayed to the coffee stem only.
- Copper oxychloride stem paint – mixed at 300 g of copper oxychloride in 1 l of water in a small bucket. A paint brush was used to apply the fungicide paint from soil level up to 50 cm on the coffee stem.
    - Frequency:
        - Copper spraying was applied once a month during rainy season and once every 3 months during dry season.
        - Copper stem paint was applied to the stem at the frequency of once in 4 months.

Table 9.1. CWD management options agreed upon and evaluated in each country (a tick [√] under each country represents options that were tried in the country).

	Agreed and tried options in each country			
Management options	DRC	Ethiopia	Tanzania	Uganda
Mulch	√	√	√	√
Herbicide		√	√	√
Ash		√	√	√
Copper stem paint	√	√	√	√
Copper stem spray	√	√	√	√
Hoe weeding	√		√	√
Slash and hand weeding		√		
Slashing only	√	√		
Pruning	√			

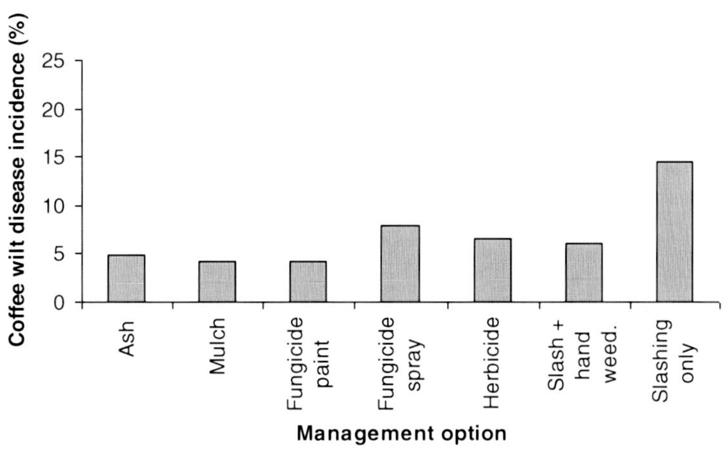

Fig. 9.1. CWD incidence on *C. arabica* in Ethiopia.

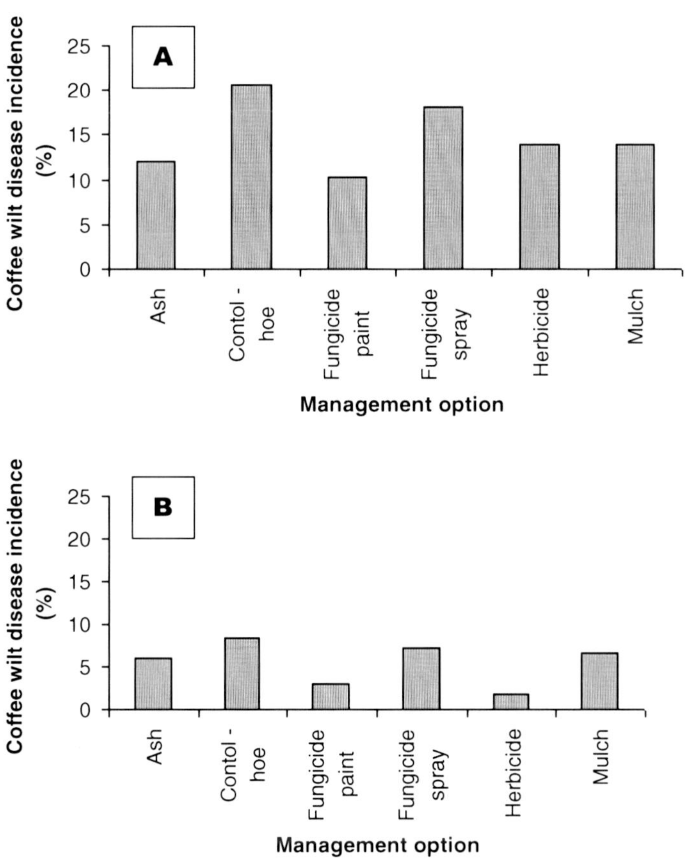

Fig. 9.2. CWD incidence on *C. canephora* in Tanzania (a) and Uganda (b).

- NPK (nitrogen–phosphorus–potassium) fertilizer was applied at the rate of 250 g per tree twice per year during the rainy season.

Some promising results: It was apparent that all management options were effective in managing CWD in Ethiopia (Fig. 9.1). In particular, the result was particularly encouraging in that the addition of hand weeding around the coffee stem as opposed to the normal practice of slashing in coffee fields significantly reduced the incidence of CWD on *C. arabica* in Ethiopia (Fig. 9.1). It was concluded that hand weeding prevents wounding, which is associated with use of a bushman's knife in Ethiopia. Farmers have since adopted hand slashing plus hand weeding, mulching and use of herbicide for the management of CWD in Ethiopia.

Studies on *Coffea robusta* in Tanzania and Uganda (Fig. 9.2a and b) did not give as marked differences as those on *C. arabica* in Ethiopia. However, copper oxychloride applied as stem paint and herbicide lowered the incidence of CWD. These studies were carried out over a 3-year period. Because they were carried out in the field under natural inoculum, there is need to carry them out for a much longer period, between 5 and 7 years.

9.3. Use of Resistant Varieties (Clones)

Use of resistant varieties is the most appropriate, efficient and economical method for the management of CWD. It is as environmentally friendly as most cultural control methods. When CWD first occurred in 1927, use of resistant varieties in combination with cultural methods reduced the impact of the disease in the 1950s, and CWD became regarded as a minor disease until its re-emergence in 1986 in DRC. Several authors have reported varietal differences in resistance to CWD and suggested the use of resistant varieties as a means of controlling CWD (Fraselle, 1950; Deassus, 1954; Bouriquet, 1959; Porteres, 1959). Cultivars of *C. canephora* (notably robusta), which were resistant, formed the basis of breeding programmes in many of the West African countries. Van der Graaff and Pieters (1978) reported that coffee lines of *C. arabica* in Ethiopia showed differences in resistance to the CWD pathogen, thus providing potential for controlling CWD using resistant varieties in arabica coffee. They suggested that resistance in *C. arabica* was quantitative in nature and horizontal, and there was no evidence of single-gene (vertical) resistance that could be readily overcome by pathogen adaptation. During the current studies under the RCWP, a series of screening trials were carried out in participating countries, which aimed at finding resistant coffee varieties and clones. In many cases, seedlings and rooted cuttings were raised from coffee bushes, which were apparently healthy in coffee fields where nearly 100% of the coffee bushes had been destroyed by CWD. In addition, seedlings and rooted cuttings were also obtained from existing coffee germplasm collections in these countries. Collections from the two sources were screened by artificial inoculation in a spore suspension of the two strains of *G. xylarioides* (from arabica and robusta coffees). Screening was also carried out in fields by assessing for over 5 years. In Uganda, 8 resistant clones were identified through artificial inoculation and field screening, which have been released for growing by farmers but are yet to be commercialized. Further resistant clones are also under multilocational field evaluation (for further details, see Chapter 9).

In Tanzania, 851 lines were screened, and 273 robusta lines have been identified to have resistance to CWD. The materials are currently being raised in a clonal mother garden for further assessment. The materials from these mother gardens will be assessed further for other agronomic traits through on-farm and national trials.

Screening studies in DRC reported that 35 coffee lines were already collected and established in two areas with CWD.

9.4. Chemical Control

CWD is a systemic disease that infects vascular bundles; therefore, controlling it with chemicals is a very difficult option. The best way of controlling CWD is by preventing infection from occurring or by using resistant varieties. However, resistant varieties are not yet available to every farmer in the affected countries. In addition, susceptible varieties or clones will still be available even after resistant varieties are available to farmers, thus warranting alternative strategies such as fungicides. The pathogen requires wounds for infection, and because of its systemic nature, one method for managing CWD is therefore to treat pruning wounds (either when changing a cycle or for rejuvenation, or during normal pruning operations). Copper-based fungicides can be used for sealing the wounds after forming a paste from the fungicide (Phiri, personal observation).

In addition to using fungicides for sealing wounds, studies on spraying coffee stems or painting the first 50 cm of a coffee stem from the ground level with a copper-based fungicide showed some promise in reducing the incidence of CWD in on-station and on-farm trials in a number of participating countries (see 9.2.5, 'Trials on Evaluation of Cultural and Chemical Control Methods for the Management of CWD in Ethiopia, Uganda, Tanzania and DRC').

Earlier studies by Gaudy (1956) showed that spraying coffee bushes with copper oxychloride was effective in controlling CWD. There is therefore potential in using copper-based fungicides, but the economics of using a fungicide will be costly.

Control of CWD with systemic fungicides can be an alternative option, but again, its economic viability is likely to be low, and the risk of contaminating the coffee beans with the fungicide should be investigated. There is also the possibility of the pathogen adapting to systemic chemicals. Systemic fungicides are already being used in the control of CLR, mostly by large-scale coffee farmers. Even if systemic fungicides are found to be uneconomical to use for controlling CWD or are found in the coffee bean, they may be the only solution for saving valuable coffee germplasm materials that have useful agronomic traits but are susceptible to CWD. Many CWD-affected countries are losing their valuable germplasm. Uganda has lost coffee germplasm due to CWD (Hakiza, personal communication). In addition, Uganda has lost *Coffea kapagota* at the Entebbe Botanical Gardens due to this disease. An effective systemic fungicide could therefore be used to protect these valuable materials. Preliminary studies with benomyl under greenhouse conditions in Uganda demonstrated that the fungicide may control CWD, but frequent drenching every month was required (Hakiza, personal communication).

9.5. Biological Control

Biological control is defined as the use of a living organism to control or manage another living organism. Natural enemies include parasites, predators, fungi, nematodes and viruses. Biological control agents (natural enemies) are currently unavailable for *G. xylarioides*. However, earlier studies by Rabechault (1954) reported four actinomycetes, one bacterium and *Corticium*, *Marasmius* and *Trichoderma* spp., which had inhibitory effects to *G. xylarioides*. Biological control could have great potential for the control of CWD as a sealant for the wounds brought about by rejuvenation (stumping), and this should be investigated further.

9.6. Integrated Crop and Pest Management

Integrated crop and pest management (ICPM) is an integrated approach to crop health management. ICPM is ideal for management of CWD and involves use of as many management methods as possible to minimize problems caused by insects, plant diseases and crop management. Management of CWD includes cultural management (prevention of injury and ensuring proper crop nutrition), management of insect pests such as stem borer and management of CWD itself. The methods include cultural, mechanical, physical, environmental, chemical and biological control. The components of ICPM must be used in a systematic way; it must include a proper monitoring programme for CWD and for insect pests and a proper identification of these so that ideal components are chosen and combined to manage the disease effectively. Use of information systems, in particular training of extensionists and farmers, and dissemination of information are crucial for the ICPM approach to manage CWD.

The main objectives of an ICPM programme are to eliminate or reduce the initial inoculum for the disease, to reduce the effectiveness of initial inoculum, to increase the resistance of the crop, to delay the onset of the disease and to slow down the secondary cycles. ICPM involves the selection and application of a harmonious range of disease-control strategies that minimize losses and maximize returns. ICPM of CWD can therefore include a combination of some of the following methods: (i) use of resistant or tolerant varieties; (ii) provision of balanced crop nutrition; (iii) use of healthy (clean) planting material; (iv) effective quarantine; (v) management of infected coffee plant materials; (vi) control of insect vectors or those that predispose the coffee tree; (vii) use of cultural practices, for example preventing tree wounding during weed management; (viii) choosing a clean field to establish coffee planting; (ix) suppression of the pathogen in infected fields; (x) prevention of the spread of disease in the field and (xi) use of pesticides. It is this approach which was advocated during the RCWP, and methods for disseminating these messages are given in more detail in Chapter 11.

9.7. Governance Issues Related to CWD and Other Disease Epidemics

A detailed study was carried out by Quinlan *et al.* (2006) that included governance issues related to CWD and other epidemic diseases. Although coffee is the most important cash crop, a major source of household income and is fundamental to livelihoods in more than 20 African countries, little attention was paid by national governments to managing the problem of CWD. It took too long from the time of initial detection, recognition of the severity of the problem and conducting of surveys. Initially, this was due to the challenge of conducting any surveillance in areas of conflict in DRC. Yet, there was also a lack of preparedness and absence of an early warning system to trigger responses in other neighbouring areas. Existing institutions dedicated to coffee research supported CWD detection, identification and management after the initial discovery of the disease. However, even after the disease was identified, the lag time in responding suggests a lack of technical and institutional capacity and a lack of an adequate strategy to deal with the problem. Inadequate budgetary support was also a big problem.

9.7.1. Preparedness before the outbreak of CWD

It is difficult to judge the preparedness of a country or countries for a disease outbreak once it has occurred. Perhaps the best measure is whether relevant institutions were in place before the outbreak occurred. Coffee research institutes were already established in all affected countries before the re-emergence of CWD. However, for a disease that spreads rapidly, the existence of coffee research institutes alone was simply not sufficient. The ongoing conflict in DRC at the time when CWD was identified in that country exacerbated the problem and is undoubtedly a major factor in the disease not receiving the attention it deserved, especially in the initial foci of infection where it could easily have been eradicated.

Initial investigations were also carried out in Uganda by a team of scientists from the then Coffee Research Centre (COREC) (now CORI) after they received reports of a disease outbreak on coffee from the Bundibujyo and Rukungiri districts, which border DRC. CWD was not confirmed during these initial investigations, and the outbreak was thought to be of relatively minor importance (Birikunzira and Lukwago, 1997) (see Chapter 3 for further details).

In 1996, DRC through ICO requested funds to confirm the identity of the disease outbreak 'before their coffee was completely destroyed'. CABI, together with scientists from DRC, plus officials from OZACAF, undertook a preliminary survey mainly in Isiro and Beni. The survey confirmed the presence of CWD, and a formal report was produced (Flood, 1996). Although there was a coffee research capability under the National Institute of Agricultural Studies and Research (INERA) prior to the outbreak of CWD, activities concerning the management and control of CWD were not initiated until the launch of the RCWP. This was largely due to the civil conflict that erupted in DRC in 1996.

There were no early warning systems in place in DRC, and although the first reports of CWD emerged in the 1970s, it was nearly 25 years before OZACAF requested the assistance from ICO in confirming a possible outbreak of CWD (Chapter 2). In contrast, the response to the outbreak of CWD in Uganda, once it had been officially confirmed, was rapid. The Ugandan government through COREC funded preliminary surveys to establish the occurrence of CWD, and other surveys were undertaken under the RCWP.

9.7.2. Initial detection of CWD resurgence

In Uganda, the CWD outbreak was first noticed by local agriculture personnel in the Bundibujyo and Rukugiri districts who reported it to COREC through their district offices. Initial visits failed to confirm the identity of the disease probably due to the high incidence of other pests and diseases. This changed dramatically when the same symptoms were reported in the Mukono district, 300 km away. The samples collected by COREC yielded *F. xylarioides* (G. Hakiza, 1995, unpublished data). Other government institutes that played an active role when the outbreak occurred were the Uganda Coffee Development Authority and the Faculty of Agriculture of Makerere University (Chapter 3). Following the launch of RCWP, the aforementioned institutes continued to collaborate in the management of and research aimed at controlling CWD in Uganda. COREC led the Ugandan team.

In DRC, the confirmation of the re-emergence of CWD was carried out by CABI (Flood, 1996) at the request of OZACAF (now Office National du Café [ONC]) through ICO. OZACAF prepared an initial report that outlined the scale of the problem and submitted it to ICO with a view to getting external funding to identify and manage the disease. ONC has been responsible for coffee extension services in DRC since 1991, but these services have not been particularly effective in rural areas in recent years due to problems associated with the deterioration of the infrastructure in the rural areas, political instability and armed conflict (ICO, 2005). This to a great degree also affected the monitoring and surveillance of CWD in the early stages of the outbreak. Partners involved in CWD at a national level include the University of Kinshasa, the National Programme of Research on Coffee, under the INERA and ONC, who are leading the programme in DRC. INERA and ONC are both under the Ministry of Agriculture.

Based on the aforementioned experiences in Uganda and DRC, there are a number of strategies that should now be put in place to avoid a recurrence of CWD or any other invasive pest/disease:

1. Legislation should provide for the destruction of infected coffee crops, together with coffee crops that may not necessarily be infected but are in the proximity of infected crops and could eventually become a secondary source of inoculum.
2. It is important that scientists and extensionists are trained in the identification of CWD even in the absence of an outbreak to reduce the possibility of incorrect diagnoses.
3. Raising awareness of the existence of CWD is of paramount importance in containing an outbreak. Given the high percentage of rural communities, a farmer-participatory approach to raising awareness is essential.
4. Communities along the borders between affected and non-affected countries should be sensitised so that they do not cross with the concerned coffee materials/products using unofficial routes.
5. Once an outbreak of CWD has been reported in a particular region, it is essential that trained scientists undertake routine surveillance, especially in border areas, so that any outbreaks can be contained rapidly and more cost-effectively using early warning systems.
6. Once an outbreak has occurred, the area must be cordoned off/quarantined immediately to prevent further spread. This could have prevented the spread of CWD in Uganda where the initial disease outbreak was on the boarder of DRC. Instead it was carried rapidly to the Mukono district, 300 km from the first disease foci, and subsequently to all coffee-growing districts.
7. All borders with affected countries should be closed for the movement of any coffee materials/products to reduce the risk of contaminated coffee material entering the country. This may involve employing extra quarantine personnel to man the borders.

9.8. Conclusion

CWD remains a huge threat to coffee production in Africa, and it is a potential threat to coffee production in the other coffee-growing continents, such as South America and Asia. However, a number of methods that can help in managing CWD are available. Notable ones are uprooting and burning CWD-infected coffee trees; prevention of tree wounding; use of protective fungicides in sealing wounds, particularly those from changing crop cycles and pruning; quarantine (very important especially for preventing intercontinent spread of CWD); biological control and use of resistant varieties. Use of systemic or curative fungicides has potential particularly in saving *in situ* coffee germplasm from decimation by CWD. The use of systemic fungicides on a commercial scale needs to be investigated further. Biological control has a very high potential, especially as a sealant for the coffee stumps after changing the coffee cycle. However, use of resistant varieties is probably the most cost-effective method for the management of CWD and is relevant to smallholder production. Resistant varieties and clones have been identified, although a lot more work needs to be done before they become available to all coffee farmers in the affected countries. It is therefore imperative that government institutions in all coffee-growing countries, particularly in Africa, take breeding for resistance as a priority for the control of CWD, and governments should provide the necessary financial and political support. Quarantine is crucial at this stage when a number of countries are sharing borders with CWD-affected countries. Monitoring and surveillance activities ought to be maintained to curb a possible spread of CWD from the affected countries. Dissemination of information and training of farmers are crucial in the process of controlling CWD and any other pest epidemics of CWD's magnitude and should continue as long as coffee is being grown. What has happened in Central Africa should be taken as a lesson-learned exercise for other coffee-growing states. Governments of CWD-affected countries were not prepared for the resurgence of CWD and did not act fast enough to curb further spread of CWD in their countries. It is a matter of time before CWD spreads to other countries. The control of CWD therefore requires joint efforts from all coffee-growing countries, not only in Africa but also in the other continents as well. They need to set aside funds for surveillance activities. For example, Rwanda, although not affected directly, was supported to carry out surveillance activities, and this would greatly help in limiting the spread and impact if there should be a CWD outbreak in the country.

In conclusion, it is important that governments are willing to share the resistant coffee germplasm with those that do not have. This is the only way of making CWD history as happened in the earlier decades of the 20th century.

Acknowledgements

This chapter is an output of Project 4 activities of the RCWP, which was funded by the Common Fund for Commodities. The ICO supervised the project, whereas CABI was the project execution agency. The authors sincerely thank all the institutions, scientists, extensionists and farmers who participated in implementing the project activities from which most of the information in this chapter have come.

References

Birikunzira, J.B. and Lukwago, G. (1997) The status of coffee wilt disease (tracheomycosis) in Uganda: a country report. In: *Proceedings of the first regional workshop on the coffee wilt disease (Tracheomycosis).* Kampala, Uganda, pp. 98.

Bouriquet, G. (1959) Plant diseases and pests in some African territories. *FAO Plant Protection Bulletin* 7, 61–63.

CABI (2006) *Crop Protection Compendium*, 2006 edn.

Clowes, M.St.J. and Hill, R.H.K. (1981) *Coffee Handbook*, 2nd edn. Zimbabwe Coffee Growers Association, Harare, Zimbabwe.

Deassus, E. (1954) La tracheomycosis de Cafeier. *Bull. Sci. Minist. Colon. Sect. Agron. Trop.* 5, 345–348.

FAO (2004) FAOSTAT data, 2004.

Flood, J. (1996) Observations on a coffee disease in North Eastern Zaire (July–August 1996) Report to OZACAF August 1996.

Flood, J. and Brayford, D. (1997) Re-emergence of *Fusarium* wilt of coffee in East and Central Africa. In: *Proceedings of the 17th International Conference on Coffee Science*. Association Scientifique Internationale du Cafe (ASIC), Paris, France, pp. 621–628.

Fraselle, J. (1950) Observations preliminaires sur une tracheomycosis de *Coffea robusta*. *Bulletin Agricole du Congo Belge* 41, 361–372.

Gaudy, M.R. (1956) Contribution du techniques, scientifique ou développement de l'agriculture en Afrique Occidentale Francaise. *Journal of the West African Science Association* 2, 172–197.

Girma, A. and Hindorf, H. (2001) Recent investigation on coffee tracheomycosis, *Gibberella xylarioides* (*Fusarium xylarioides*) in Ethiopia. In: *Proceedings of the 19th International Scientific Conference on Coffee Science (ASIC)*. Trieste, Italy, pp. 1246–1252.

Girma, A., Hulluka, M. and Hindorf, H. (2001) Incidence of tracheomycosis, *Gibberella xylarioides* (*Fusarium xylarioides*), on arabica coffee in Ethiopia. *Zeitschrift für Pflanzenkrankheiten und Pflanzenschutz* 108, 136–142.

ICO (2005) Exports by exporting countries to all destinations. Available at: http://www.ico.org/trade/m1.htm.

Jacques-Felix, H. (1954) La carbuncularicse. *Bull. Sci. Minist. Colon. Sect. Agron. Trop.* 5, 296–344.

Krantz, J. (1962) Coffee diseases in Guinea. *FAO Plant Protection Bulletin* 10, 107–110.

Kranz, J. and Mogk, M. (1973) *Gibberella xylarioides* Heim et Saccas on arabica coffee in Ethiopia. *Phytopathologische Zeitschrift* 78, 365–366.

Oduor, G. and Simons, S.A. (2003) Biological control in IPM for coffee. In: Neuenschwander, P., Borgemeister, C. and Langewald, J. (eds.) *Biological Control in IPM Systems in Africa*. CAB International, Wallingford, UK.

Oduor, G., Simons, S., Phiri, N., Njuki, J., Poole, J., Pinard, F., Kyetere, D., Hakiza, G., Musoli, P., Lukwago, G., Abebe, M., Tesfaye, A., Kilambo, D., Asiimwe, T. and Munyankere, P. (2003) Surveys to assess the extent and impact of Coffee Wilt Disease in East and Central Africa. Final Technical Report EU contract No. ASA-RSP/CV-006. CAB International, Egham, UK.

OZACAF (1995) Project de lutte contre la tracheomycose du Cafeier au Zaire. Report for Office Zairois du Cafe (OZACAF) on coffee wilt in Zaire.

Porteres, R. (1959) Valeur agronomique des Cafeiers des types Kouilou et Robusta cultivars en Cote d'Ivoire. *Café Cacao* 3, 3–13.

Quinlan, M.M., Phiri, N., Zhang, F. and Wang, X. (2006) Foresight, Infectious Diseases: Preparing for the future. The Influence of Culture and Governance on Detection, identification and Monitoring of Plant Disease, A comparative assessment of the United Kingdom, China and Sub-Saharan Africa. Office of Science and Innovation, London, United Kingdom. Available at: http://www.foresight.gov.uk/Infectious%20Diseases/d4_1.pdf.

Rabechault, H. (1954) Sur quelques facteurs de resistance du Cafier a la Tracheomycose. *Bull. Sci. Minist. Colon. Sect. Agron. Trop.* 5, 292–295.

Rice, R.A. and Ward, J.R. (1996) Coffee, Conservation and Commerce in the Western Hemisphere. In: Smithsonian Migrator Bird Centre and Natural Resources Defence Council, Washington, DC. p 4.

Saccas, A.M. (1956) Recherches experimentales sur la tracheomycose des cafeiers en Oubangui-Chari. *Agronomia Tropical Nogent,* 11, 7–38.

Van der Graaff, N. and Pieters, R. (1978) Resistance levels in *Coffea arabica* to *Gibberella xylarioides* and distribution pattern of the disease. *Netherlands Journal of Plant Pathology* 84, 117–120.

Wrigley, G. (1988) *Coffee.* Longman Press, London, UK.

10 Breeding for Resistance Against Coffee Wilt Disease

P.C. Musoli,[1] A. Girma,[2] G.J. Hakiza,[1] A. Kangire,[1] F. Pinard,[3] C. Agwanda[4] and D. Bieysse[3]

[1]*Coffee Research Centre, PO Box 185, Mukono, Uganda*
e-mail: cori@africaonline.co.ug
[2]*Jimma Agricultural Research Center, Ethiopian Institute of Agricultural Research, PO Box 192, Jimma, Ethiopia*
e-mail: girma.adugna@yahoo.com
[3]*Centre de Coopération Internationale en Recherche Agronomique pour le Développement (CIRAD), TA41/K, Campus International Baillarguet, 34398, Montpellier, Cedex 5, France*
[4]*CABI Africa, UN Avenue, ICRAF Complex, PO Box 633-00621, Nairobi, Kenya*
e-mail: c.agwanda@cabi.org

10.1. Introduction

Coffee wilt disease (CWD) is very destructive to coffee trees, which leads to enormous loss of investment in coffee farming. The disease was first reported in 1927 on *Coffea liberica* var. *dewevrei*, formerly *C. liberica* type excelsa (Davis *et al.*, 2006), in the Central African Republic (CAR) (Figueres, 1940). It progressively destroyed this crop during the 1930s to the 1950s in other Central African countries, particularly in Cameroon (Guillemat, 1946; Fraselle, 1950; Saccas, 1951; Muller, 1997). During the same period, it destroyed *Coffea canephora* in the Ivory Coast, causing serious losses and disappearance of some local varieties (Delassus, 1954). In contrast, several varieties of *C. canephora* imported from the Democratic Republic of Congo (DRC) between 1914 and 1933 exhibited some level of field resistance, which was later confirmed through artificial inoculation (Meiffren, 1961). Meiffren (1961) also reported apparent differences for the same materials planted in different areas of the region, i.e. certain *C. liberica* and *C. canephora* varieties showing resistance in Ivory Coast were completely susceptible in CAR, suggesting the resistance was either being influenced by environmental conditions or there were different physiological races of the pathogen in different localities of this region. Fraselle (1950) reported CWD attack on *C. canephora* at Yangambi in DRC in 1948, and subsequently, the disease became a potential problem for the country. In 1957, similar symptoms were reported on *Coffea arabica* in Ethiopia (Lejeune, 1958), and later, Kranz and Mogk (1973) confirmed that the disease on *C. arabica* was also caused by *Gibberella xylarioides*. Pieters and Van der Graaff (1980) reported that CWD was endemic in all coffee-growing areas of Ethiopia. Van der Graaff and Pieters (1978) reported resistance among *C. arabica* lines in fields, which was later confirmed after artificial inoculation (Pieters and Van der Graaff, 1980).

As CWD threatened the coffee industry throughout Africa, affected countries decided in 1956 to implement systematic elimination of all affected plants over large areas and to search for resistance both in wild and cultivated varieties. Following this initiative, *C. canephora*-resistant varieties identified in DRC were used for replanting within DRC and Ivory Coast (Saccas, 1956; Meiffren, 1961). In Cameroon, the disease was eliminated by rigorous systematic uprooting (Muller, 1997) of the *Coffea excelsa* var. *dewevri* plantations. These strategies proved to be successful as the disease had declined drastically by the end of the 1950s, and it eventually disappeared from Cameroon and Ivory Coast, and probably DRC and CAR. However, the disease continued affecting *C. arabica* in Ethiopia, and it is doubtful if any of these strategies were applied there. In 1986, new large-scale outbreaks of CWD were reported on *C. canephora* in the north-east of DRC (Flood and Brayford, 1997), from where it spread rapidly into Uganda (1993) and north-west Tanzania (1996). Because the disease appeared in these countries for the first time, there were no resistant varieties available for replanting in infected areas and all available commercial varieties were susceptible to CWD, so replanting with these materials in contaminated soils was not an option. In addition, there were no effective phytosanitory management practices. Thus, following the successful use of resistance in Ivory Coast and the CAR, in Uganda, a breeding programme was initiated at the Coffee Research Centre (COREC) (now CORI) which aimed at developing resistant germplasm for managing the disease. Similar breeding programmes were initiated by TaCRI in Tanzania and the University of Kinshasa in DRC. The national breeding programmes in the respective countries were implemented independently, although the programmes in DRC and Uganda had a strong linkage through their collaboration with Centre de Coopération Internationale en Recherche Agronomique pour le Développement (CIRAD) in France.

10.2. Potential Sources of Resistance

Intra- and interspecific differences among and between coffee species respectively provide potential genetic variability, which is exploited for resistance against CWD. Intraspecific variability is the best and easiest to exploit since resistant individuals are easily released as new varieties without undergoing hybridization, provided they posses other agronomic traits such as being high yielding; having resistance to other major diseases, mainly leaf rust and red blister disease and coffee berry disease (arabica only) and having good market qualities (big beans and good cup qualities). Where necessary, intraspecific hybridization is easier to carry out, and it is usually more successful.

C. canephora is particularly genetically variable (Plate 26), and the variability is very diverse even among genotypes from the same locality (Musoli, 2007) or members of the same progenies, mainly because it is out breeding and its wide geographical distribution (Leakey, 1970). Musoli (2007) found that most of the genetic diversity among *C. canephora* populations at molecular level was attributed to variations (heterozygosity) within individuals. Similar studies carried out in DRC revealed diverse variability among *C. canephora* populations in that country. *C. arabica* is relatively less genetically diverse, but nevertheless, the available diversity is high enough to be exploited for resistance against CWD.

Open pollinated seedlings of different *C. canephora* populations, which included two cultivated distinct morphological types ('erecta' and 'nganda'; Plate 27) plus wild populations from Kibale and Itwara forests and a feral population from Kalangala, an isolated island in Lake Victoria, were assessed for CWD resistance through artificial inoculation in a screen house at COREC. These studies showed significant (P = 0.001) genetic differences between the populations for CWD resistance. The disease (measured by plant mortality) progressed at different rates and to different final levels for the different populations (Fig. 10.1). This illustrates the usefulness of a diverse germplasm population when sourcing for resistance against CWD. Artificial inoculation of different genotypes of *C. canephora* in DRC and Tanzania also showed varying levels of resistance among the genotypes. Similar studies carried out on *C. arabica* in Ethiopia revealed differential reactions to CWD by genotypes from different localities (Chapter 8). In Uganda, exploitation of intraspecific variability in *C. canephora* led to selection of eight high-performing, CWD-resistant clones,

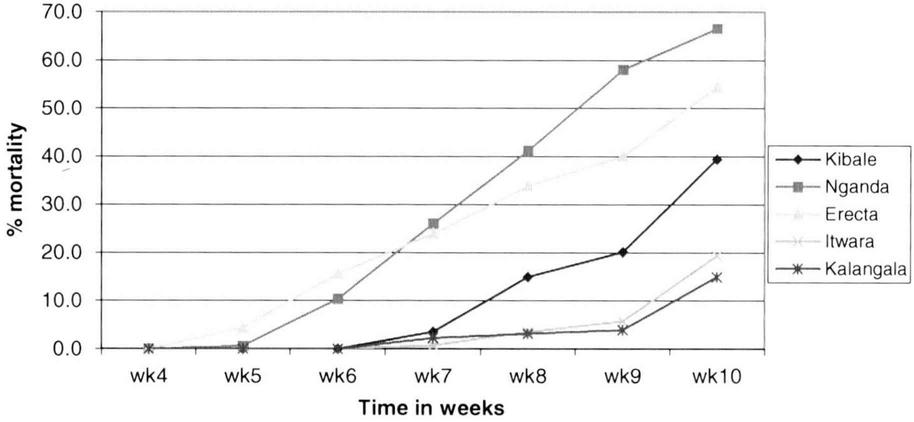

Fig. 10.1. Progression of CWD among open pollinated progenies of different *C. canephora* populations.

which also have good bean and cup qualities and resistance to red blister disease and leaf rust. These have been recommended for commercial cultivation.

In some parts of Uganda where *C. canephora* and *C. arabica* are cultivated side by side, CWD affected only the former species. Similar responses have been reported in Ethiopia, Tanzania and DRC. The difference observed between the two species in Uganda was confirmed through artificial inoculation of seedlings of the two species in the screen house (Musoli *et al.*, 2001). Interspecific hybrids (F1, backcross 1 and backcross 2; arabica as a recurrent parent) of *C. canephora* and *C. arabica* (arabusta) grown close to *C. canephora* fields affected by CWD have remained unaffected since 2001, when the disease was first observed on *C. canephora*. The resistance of the arabusta hybrids was also confirmed through artificial inoculation of clones of their F1 progenies and F2 progenies at COREC. This implies that the resistance to CWD in *C. arabica* is dominant, and it can be introgressed into *C. canephora* through interspecific hybridization. However, deriving *C. canephora*-resistant varieties through arabusta interspecific hybrids needs several backcrosses to the *C. canephora* recurrent parents This backcrossing is complex because the two species have different ploidy levels and the crosses normally involves *C. canephora* artificially generated tetraploids.

Some level of host specificity to contemporary strains of *G. xylarioides* was noted through host–pathogen interaction studies. The strain currently affecting *C. canephora* in Uganda, DRC and Tanzania is specific to this species. Artificial cross inoculations carried out on young *C. canephora* and *C. liberica* half sib progenies in growth rooms showed that *G. xylarioides* isolate (CAB003), which was obtained from *C. canephora*, causes severe mortality only on *C. canephora* (Fig. 10.2a) (Musoli, 2007). Similar studies showed that the strain affecting *C. arabica* in Ethiopia is specific to this species (Chapter 8). In mixed coffee cultures of *C. canephora and C. arabica* within Ethiopia, the pathogen strains infect only *C. arabica*. Artificial inoculation of *C. canephora* under controlled growth room conditions using isolate CAB007 collected from *C. arabica* in Ethiopia showed that this strain induces early symptoms of CWD on *C. canephora*, but it is not fatal (Fig. 10.2b). The interaction of a historical strain, DSMZ62457, isolated in the previous epidemic during the 1950s on *C. canephora* in CAR and evaluated through artificial inoculations in the growth room gave mixed reactions. Although Musoli (2007) showed that this strain was an aggressive pathogen of *C. liberica* and moderate on *C. canephora* (Fig. 10.2b), other studies showed that *C. liberica* isolate DSMZ62457 was pathogenic to *C. canephora*, *C. liberica* and *C. arabica* and that other isolates from *C. arabica* and *C. canephora* induced some symptoms on *C. liberica*. Therefore, further studies are required to validate these findings. Field reports from DRC and Uganda suggest that certain *C. liberica* spp. are susceptible to the *C. canephora* isolate within these countries. These observations thus exclude *C. liberica* from being a source of resistance to CWD for introgression into arabica and robusta. Given such host–pathogen specificity, varieties with durable resistance, which can be used across the entire African region, irrespective of the prevailing pathogen strain, can be derived through Arabusta interspecific hybridization.

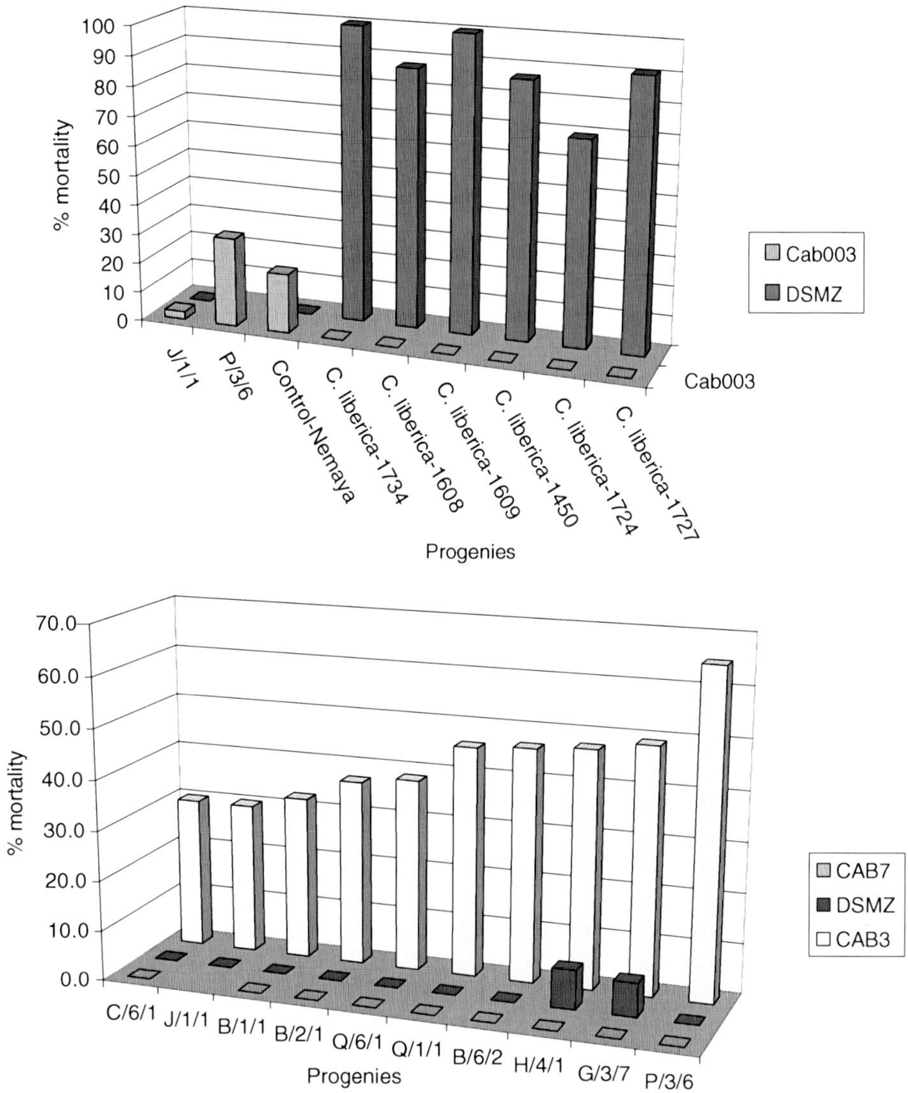

Fig. 10.2. (a) Mortality among *C. canephora* and *C. liberica* var. *dewevrei* seedlings inoculated with *G. xylarioides* strains CAB003 (*C. canephora*) and DSM62457 (*C. liberica*). Nemaya is *C. canephora* progeny given for this purpose by CIRAD. J/1/1 and P/3/6 are *C. canephora* progenies obtained from Uganda. (b) Mortality of *C. canephora* seedlings inoculated with *G. xylarioides* strains CAB003 (*C. canephora*), CAB007 (*C. arabica*) and DSMZ62457 (*C. liberica*).

10.3. Assessment of Resistance

Use of a quick and effective procedure is crucial to the successful identification of reliable resistance. Where breeding for resistance against CWD is being initiated for the first time, it is necessary to have the basic skills and knowledge, which can lead to successful screening of the germplasm and identifying genuinely resistant genotypes. This will include having easy but effective protocols for testing and quantifying the resistance. The screening protocols involve assessing genotypes for resistance under natural and artificial infections. There are almost no special manipulations involved when assessing coffee plants for resistance in naturally infected gardens, and limited variation in methodologies is anticipated, except perhaps in the quantification of resistance. There are more methodological variations when assessing genotypes for CWD resistance in artificial conditions, perhaps due to variation of costs required by different methods and the facilities available for the studies. Thus, a greater part of this section deals with methods involved in artificial inoculation.

10.3.1. Inoculation methods

Different inoculation procedures have been used by different scientists for artificially inoculating coffee plants when testing them for CWD resistance (Meiffren, 1961; Van der Graaff and Pieters, 1978; Bieysse, 2005; Musoli *et al.*, 2001; Musoli, 2005). In Uganda, a number of artificial inoculation methods were evaluated to determine their efficacy when testing for CWD resistance in *C. canephora* (Hakiza *et al.*, 2002). The inoculation methods evaluated include

- Root dip method: involved inoculation by dipping the entire root system of the plants into a standard conidia inoculum of *G. xylarioides*;
- Stem wounding: involved introducing a standard inoculum through artificial wounds on the stems of the study plants;
- Soil infection: involved planting test plants in soil contaminated with the CWD pathogen;
- Soil drenching without root wounding: involved drenching the soils where test plants are growing with a standard inoculum;
- Soil drenching with root wounding: involved drenching the soils where the test plants are growing with a standard inoculum but after wounding the roots while the plants remain *in situ*; and
- Leaf wounding: involved applying the inoculum on to the leaves through artificially made wounds.

All the inoculations were carried out using inoculum derived from a single conidium obtained from the same host plant and standardized at 1.3×10^6 conidia per millilitre of water. The inoculum was applied on 6- to 8-month-old seedlings or cuttings of known *C. canephora* genotypes (Plate 25). Plants inoculated by root dipping developed the disease symptoms earlier than plants inoculated by other methods and had a higher incidence of diseased plants, and there was clear contrast between susceptible and resistant genotypes. This method was therefore adopted for large-scale germplasm screening by scientists in Uganda and Tanzania. However, root dipping required a lot more inputs (polythene pots, soils and manpower) than the other methods and therefore is more costly. It was also suspected that some of the plants infected by this method could have developed the disease because of extra stress resulting from root damage incurred when stripping off soils from roots in preparation for dipping. At CIRAD, where labour is costly, stem nicking was adopted. This method can also clearly enable differentiation of resistant and susceptible genotypes. Stem nicking was also adopted for germplasm screening in Ethiopia, as it is considered to be less expensive although the disease levels among plants inoculated by root dip were always higher.

10.3.2. Inoculum concentration

At COREC, it was thought that a high inoculum concentration such as the 1.3×10^6 used by scientists during the past epidemics in Central and West African countries for routine germplasm screening could kill plants with moderate resistance, leading to discarding of useful material. Conversely, lower inoculum concentrations may not be effective enough to select plants that are resistant enough to the disease pressure in fields over long periods, as in the case where the variety is planted in heavily infested gardens. To test the optimum concentration, 6- to 8-month-old *C. canephora* seedlings were artificially inoculated by dipping their entire root system into inoculum (derived from a single conidium) at concentrations of 1.3×10^1, 1.3×10^2, 1.3×10^3, 1.3×10^4, 1.3×10^5, 1.3×10^6, 1.95×10^6 and 2.6×10^6 conidia per millilitre of water. Results of these tests revealed that infection occurs even at the lowest concentration (1.3×10^1), but this inoculum concentration was associated with long incubation periods (for symptom development) and the lowest disease incidence among the inoculated plants (Fig. 10.3). Increase in

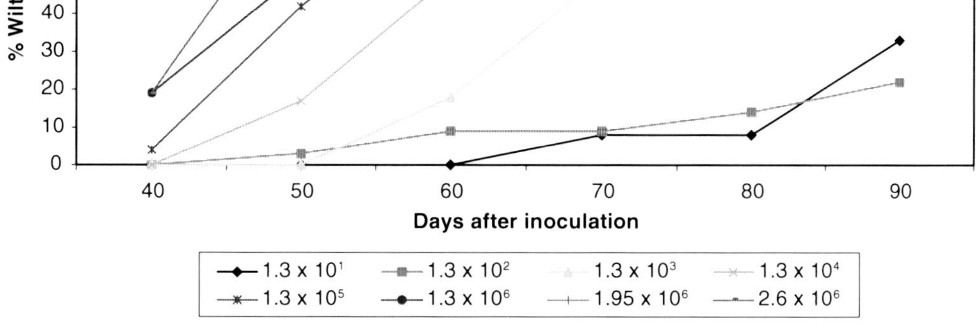

Fig. 10.3. Effect of inoculum concentration on incidence of CWD on artificially inoculated open pollinated seedlings of *C. canephora*.

inoculum concentration resulted in reduced incubation times and increased wilt incidence. It was decided that concentrations of 1.3×10^3, 1.3×10^4 and 1.3×10^5 conidia per millilitre should be adopted for routinely screening for resistance against CWD.

10.3.3. Duration of exposure to the inoculum for root dipping

To maximize the number of plants inoculated by root dipping, plants were assessed after various times of exposure to the inoculum standardized at 1.3×10^6 conidia per millilitre of water. The exposure times tested included immediate root dip (which involved withdrawal of the plants immediately after their roots had been dipped into the inoculum), 20 min of exposure (which involved withdrawal of plants after 20 min in the inoculum), 40 min of exposure, 60 min of exposure, 4 h of exposure, 8 h of exposure, 12 h of exposure and 24 h of exposure. *C. canephora* seedlings of 6–8 months were used in these tests, and the tests were all conducted twice. All inoculated seedlings developed CWD symptoms irrespective of duration of the exposure time. The differences in incidence of diseased plants, especially during the first 40 days after inoculation, were insignificant. These results showed no advantage in leaving plants in the inoculum for more than 20 min since shorter exposure periods allow more plants to be inoculated within a given period.

10.3.4. Quantifying CWD resistance in artificial inoculation

An effective and reliable method of quantifying resistance was necessary for comparison of results among experiments and among scientists, and it was necessary for the selection of genuinely resistant genotypes, irrespective of whether the evaluation was carried out on mature plants in the field or young plants in the screen house. The method adopted will depend on the purpose of the study. Where the study aims at determining relative resistance between progenies or clones, resistance can be assessed as numbers of infected plants compared to those uninfected. The numbers can be expressed as percentage infection. Relative resistance between progenies and/or clones can also be expressed on a disease symptom severity scale. In Uganda, the plants studied in artificial inoculations were commonly assessed on a scale of 1 to 5, where 1 = no disease, 2 = curling leaves and stunted growth, 3 = leaf wilting and yellowing, 4 = leaf necrosis, leaf wilting, and abscission and 5 = plants are dead. Mature plants studied in fields were also assessed on a scale of 1 to 5, but the quantification of the disease levels in the field was slightly different. In field assessment (mature coffee trees) 1 = no disease, 2 = 1%–25% defoliation, 3 = 26%–50% defoliation, 4 = 51%–75% defoliation and 5 = 76%–100% defoliation. Plants scored as level 5 are normally considered dead. The studies carried out at COREC (Uganda) and CIRAD showed that coffee trees, even of the same genotype and in heavily infected gardens, get affected at

Table 10.1. Correlation of field resistance of *C. canephora* clones to CWD measured by different traits.

	Disease severity (1–5 scale)	Disease period	Sqr. AUDPC
% plant mortality	0.997 (<0.0001)	–0.446 (0.064)	0.935 (<0.0001)
Disease severity (1–5 scale)		–0.454 (0.059)	0.931 (<0.0001)
Disease period			–0.281 (0.274)

Sqr. AUDPC is coefficient of correlations performed on the square root of AUDPC. Figures in parentheses are probability values. AUDPC = area under disease progress curve.

different rates (Section 10.3.5), and that the time lapse between the first observed symptoms and the death of trees varies between genotypes. Thus, there was an attempt to quantify relative resistance among *C. canephora* clones using duration of the disease period (average time taken by plants from appearance of first symptoms to death). However, this quantification did not give consistent results because some moderately resistant clones had a shorter disease period than some highly susceptible clones and vice versa, and its correlations with other traits were not significant (Table 10.1).

Where the aim was to identify genotypes with total resistance, plants are classified as either diseased or not diseased. Since all plants that develop CWD symptoms eventually die, only plants without the symptoms after a long period of infection (6 months for plants in artificial inoculations and not less than 5 years for plants evaluated in heavily infested fields) were considered resistant. Where necessary, these plants can be re-inoculated and assessed again for another 6 months to ascertain their resistance. The plants that remained healthy after the re-inoculation were considered to have complete resistance, and such plants were planted in mother gardens for cloning and further assessments.

10.3.5. Evaluation for CWD resistance in the field

Studies carried out on mature trees of 20 *C. canephora* clones in a field at COREC revealed variable responses of the clones to CWD infection (Fig. 10.4). The tree mortality for different clones was at varying levels when assessment started, and it progressed at varying rates to varying final levels. This showed that the clones had variable levels of resistance and the disease resistance is most likely imparted by many genes, which are variably available in the clones. Clone J/1/1, which did not succumb to CWD throughout the assessment period, and clones Q/3/4, R/1/4 and 1s/3, whose disease levels were comparatively very low, were considered resistant. Clone 1s/2 appeared to be resistant until April 2002, when its trees started dying in high numbers. This shows that some genotypes require higher inoculum concen-

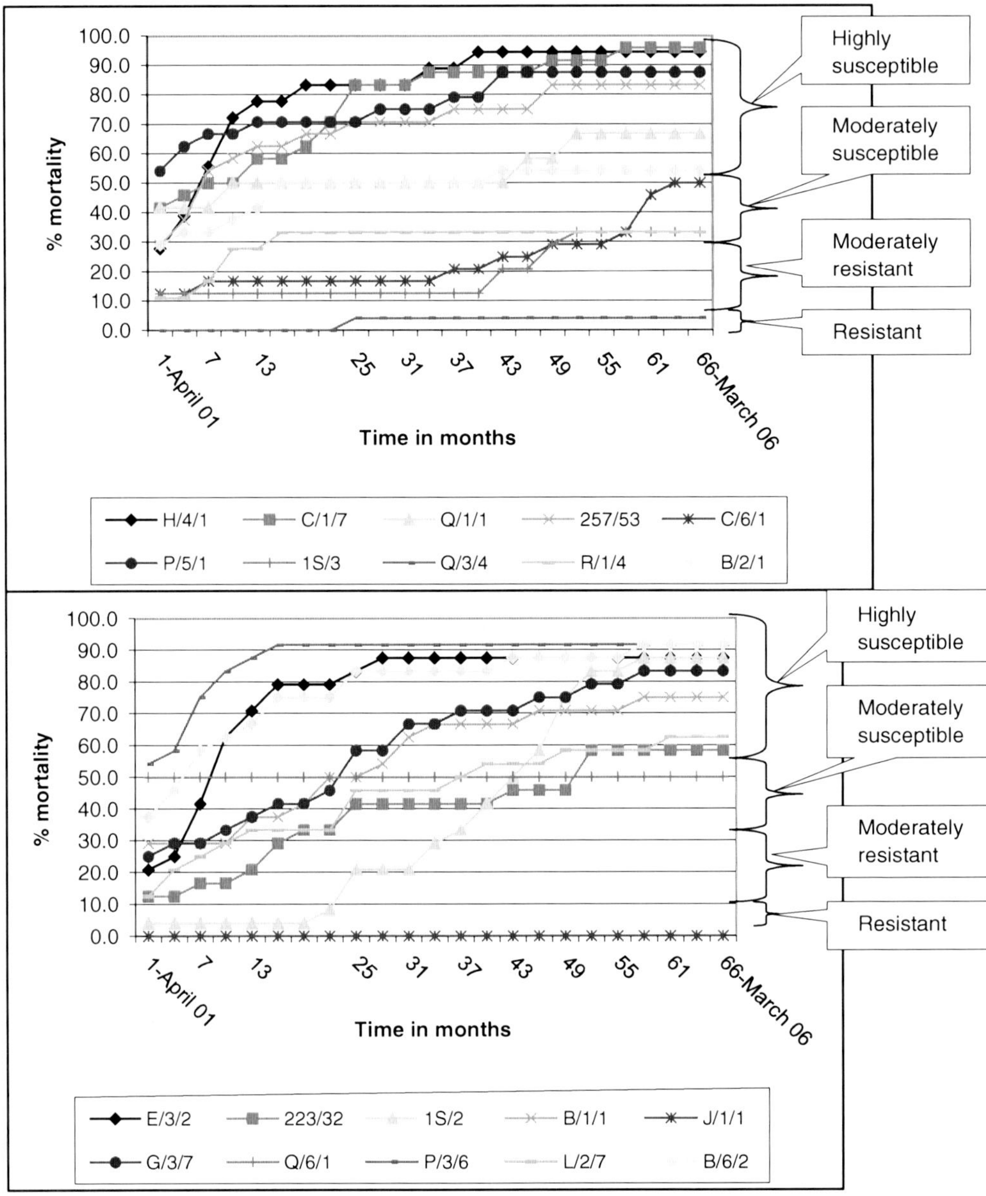

Fig. 10.4. Tree mortality of 20 *C. canephora* clones evaluated for CWD resistance in a field at Kituza.

trations for infection, and thus, there is need to assess the plants for many years, to allow the inoculum build up, when screening them for resistance in the field. Analysis of variance revealed significant genetic differences between *C. canephora* clones for resistance to CWD in the field.

10.3.6. Comparison of resistance in the field and in artificial inoculations

Rooted cuttings and open pollinated seedlings of some of the clones studied for field resistance were studied for resistance in artificial inoculations in a screen house at COREC, and their final per cent tree mortality were correlated with the final per cent tree mortality in the field. It was observed that even clone J/1/1, which had no diseased trees in the field, had 15% of its cuttings killed in artificial inoculation, but it was among the least affected (Table 10.2). Clone B/2/1 had 54.2% mortality in the field but did not have any dead cuttings after artificial inoculation. This clone could be having differential reactions to infection under the different conditions. However, there was an overall significant correlation ($P = 0.002$) between mortality in the field and mortality of cuttings in the screen house (Table 10.3). This indicates that resistance detected in artificial inoculation is depicted in the field, and therefore, the artificial inoculation is a good protocol for screening germplasm for CWD resistance. Where correlations between mortality among open pollinated progenies in the screen house and mortality of their parents in the field and in the artificial inoculations (rooted cuttings) were not significant,

Table 10.2. Mortality of *C. canephora* clones in the field and their rooted cuttings and progenies in artificial inoculation.

Clone	Field mortality	Rooted cuttings	Open pollinated progenies
J/1/1	0.0a	15.0b	
Q/3/4	4.2b	20.0b	
1S/3	33.3c		35.0abcd
R/1/4	33.3c	44.4bc	35.0abcd
C/6/1	50.0cd		65.0def
Q/6/1	50.0cd		80.0fg
B/2/1	54.2cd	0.0a	10.0a
223/32	58.3cde	90.0d	65.0def
L/2/7	62.5def	50.0bcd	25.0abc
Q/1/1	66.7defg		53.0cde
B/1/1	75.0defgh		85.0fg
257/53	83.3efgh	25.0b	40.0bcde
G/3/7	83.33efgh	80.0cd	50.0cde
P/5/1	87.5fgh		
E/3/2	87.5fgh		25.0abc
1S/2	87.5fgh	100e	20.0ab
P/3/6	91.7gh		69.0efg
B/6/2	91.7gh		68.0efg
H/4/1	94.4gh	70.0cd	60.0def
C/1/7	95.8h	63.6cd	95.0g

Means separated by Student Newman–Keuls mean separation test. Figures represented by different letters are significantly variable.

Table 10.3. Correlations of *C. canephora* resistance to CWD in field and artificial inoculations.

	Cuttings	Progeny	Progeny ($E4_1$)
Clone in field	0.965 (0.002) 6	0.132 (0.698) 11	0.708 (0.010) 12
Rooted cuttings		0.791 (0.209) 4	0.879 (0.121) 4
Open pollinated progeny			0.099 (0.798) 9

Figures in parentheses are probability values; $E4_1$ and $E4_2$ are tests 1 and 2 within experiment 4, respectively. The number below each correlation coefficient shows data pairs used to generate the coefficient.

then the response of open pollinated seedlings does not effectively represent CWD resistance of their parents.

10.4. Inheritance of CWD Resistance

Understanding the type of resistance and inheritance of the resistance was considered necessary for breeding CWD-resistant varieties, which posses other commercial attributes. Analysis of variance performed on the disease symptom severity data of clones in the field, rooted cuttings and open pollinated progenies in artificial inoculation found highly significant ($P < 0.0001$) genetic differences between the clones, rooted cuttings and open pollinated progenies (Table 10.2). These results and the responses of the different *C. canephora* populations (Fig. 10.1) indicated that CWD resistance in *C. canephora* is quantitative; thus, it is controlled by many genes that are variably distributed among the genotypes and populations.

Inheritance of CWD resistance in *C. canephora* was calculated at 50%–65% plant mortality using disease data recorded on mature trees of full sib progenies. The data were analysed using diogene quantitative genetics software (Baradat and Labbé, 1995) using Garretsen and Keuls random model adapted for incomplete/half diallel (Keuls and Garretsen, 1977). This model was able to calculate variance components and narrow (h^2_n) and broad (h^2_b) sense heritabilities. The broad sense heritability calculated from this analysis was moderate (0.329), and the narrow sense heritability was low (0.112). This shows that CWD resistance is heritable, but its transmission from parents to progenies is only about 33%; therefore, the genetic gains of choosing a progeny of resistant parents as source of planting materials for a production garden are low. Commercial CWD-resistant robusta coffee varieties should therefore be propagated vegetatively to retain the resistance. It should however be noted that resistant parents of the progenies used in this study were partially resistant, and therefore, analyses of data on progenies derived from crosses of completely resistant and highly susceptibility parents are needed to validate the current observations.

10.5. A Perspective for Developing CWD-Resistant Varieties

Basing on the information above, a practical perspective for successfully developing CWD-resistant varieties is given below. The perspective outlined is for *C. canephora*, and it might not be suitable for *C. arabica* because the two species have different reproductive systems and the interaction of *C. arabica* with the *G. xylarioides* strain is different. We have not attempted to give a possible breeding strategy for developing CWD-resistant *C. arabica* commercial varieties because we lack basic information on type of resistance and heritability of the resistance in this species. A schematic representation of the breeding outline is given in Fig. 10.5.

10.5.1. En masse germplasm screening

As already indicated, there is variability among *C. canephora* genotypes and populations for CWD resistance. Thus, en masse screening of seedlings and clones derived from different populations through artificial inoculation leads to identification of CWD-resistant genotypes quickly. In Uganda, 1519 CWD-resistant genotypes have so far been identified through en masse screening of coffee seedlings and cuttings derived from germplasm available in conservation plots at the research institutes and from coffee trees surviving in wilt-devastated gardens. The conservation plots at Kawanda Agricultural Research Institute were the immediate and most available source and therefore provided the biggest amount of seedlings and cuttings so far screened. Open pollinated seedlings and rooted cuttings were raised from various genotypes in these collections at different dates following routine nursery procedures of raising coffee seedlings and rooted cuttings (MAAIF, 1995). The seedlings and cuttings, when at 6–8 months old, were inoculated with field isolates of the pathogen prepared from a specimen obtained from a commercial *C. canephora* clone 257/53. The plants were inoculated using the root dip method and a standardized inoculum (1.3×10^6 conidia per millilitre). All inoculated plants were incubated at room conditions in the screen house as they were monitored for CWD symptoms (Plate 25). At the end of each trial, any healthy-looking plants were re-inoculated 6 months after the first inoculation. These re-inoculated plants were again incubated in the screen house and monitored for the disease symptoms. Again after another 6 months, healthy-looking plants were selected for another inoculation. Survivors of these re-inoculations were considered resistant and potential clones for future varieties. All the resistant genotypes were then planted in mother gardens and each of them cloned through cuttings to raise clones for establishing field evaluation trials (preliminary variety evaluation trials).

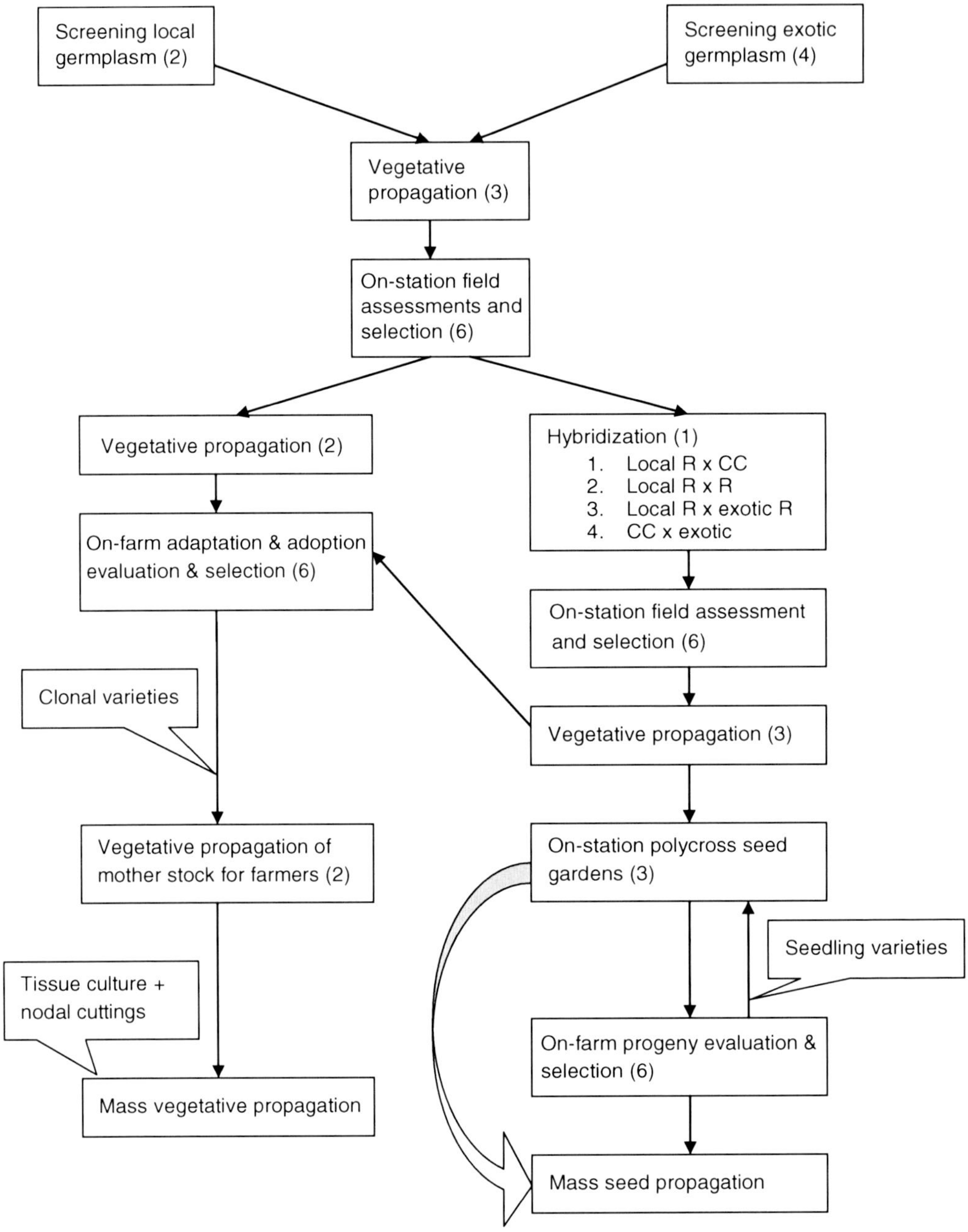

Fig. 10.5. Proposed scheme for developing *C. canephora* varieties resistant to CWD. Figures in parentheses are the minimum time in years a particular breeding stage could take. R = resistant, CC = current commercial varieties.

Open pollinated seeds and semihard wood cuttings were obtained from coffee trees surviving in wilt-devastated gardens. Seedlings and rooted cuttings were, respectively, raised from these materials for inoculation and selection of resistant genotypes following the procedure described above.

It was observed that a very small proportion (3%–15%) of the plants raised from the on-station germplasm plots survived the first inoculation. The rate of survival varied between progenies (Fig. 10.2), perhaps due to relative resistance of their parents. However, the proportion of survivors after the first inoculation from among seedlings and cuttings raised from genotypes surviving in wilt-devastated gardens was relatively higher (15%–20%). This implies that mother plants surviving in wilt-devastated gardens have some degree of resistance, although this resistance might not be strong enough to withstand the inoculum pressure given in the screen house. It has also been observed that in the re-inoculation, about 50%–70% of the re-inoculated plants die.

In Tanzania, more than 270 resistant genotypes have been identified through mass screening of germplasm. The resistant genotypes have been planted in mother gardens for vegetative multiplication.

10.5.2. Field germplasm screening

Germplasm screening can also be carried out in field observation plots. However, field evaluations tend to handle relatively fewer plants, and the assessment is prolonged over many years, yet one cannot ascertain that resistant genotypes selected through field observations can withstand high disease pressures under different agroecological conditions unless the trials are validated under a range of agroecological conditions. At COREC, clones in a field trial were assessed for CWD resistance starting 1999 to 2006. It was noted from this field trial that clones respond to the disease differently and one clone (J/1/1) was still resistant by November 2007 (Table 10.2). However, rooted cuttings raised from the resistant clone, when tested for resistance in the screen house, succumbed to the disease although the level of disease among its cuttings was lower than that among cuttings of the other clones within this trial.

10.5.3. Multiplication of the CWD-resistant varieties

Because C. *canephora* is predominantly an out-crossing crop, it is anticipated that all the individuals are heterozygous, and therefore, the CWD-resistant genotypes are expected to segregate for most traits including the resistance to CWD. It is therefore recommended that these genotypes are propagated entirely using vegetative methods, particularly rooted nodal cuttings or tissue culture. At the COREC, the former is being used for propagating materials earmarked for both further evaluation and farmers' use whereas the latter exclusively dedicated to raising planting materials for farmers' use from somatic embryogenesis.

10.5.4. Field evaluation of the resistant varieties

It is obvious that not all the CWD-resistant genotypes have all desired qualities of commercial varieties. Therefore, the CWD-resistant genotypes must undergo preliminary field evaluation, where they are evaluated for yield, resistance to other diseases (coffee leaf rust [CLR] and red blister disease in Uganda) and liquor and physical bean qualities. Because this initial evaluation involve many genotypes, it is ideal to first have a simple trial where all the available genotypes are assessed and genotypes with higher potential are selected for further evaluation in different agroecological localities. As of November 2007, 1519 CWD-resistant clones had been identified in Uganda, and these have been planted in single rows of six trees in an on-station trial at Kituza. To date, 25 superior clones have been selected from among 167 clones that have undergone the initial on-station evaluation. The 25 clones are being multiplied through nodal cuttings for further evaluation in different agroecological areas so as to validate their performance and adaptation to different localities. Good yielding (at least 2 t of green beans) clones that are resistant to CLR and red blister disease and have good liquor and physical bean qualities will be recommended for farmers' use, either across the entire country or for particular agroecological localities, depending on the performance of individual clones in the multilocation trials.

10.5.5. Hybridization for CWD resistance

Although CWD takes precedence at the screening stage, yield, quality and resistance to CLR, red blister disease and other major coffee diseases are important traits for commercial *C. canephora* varieties, particularly in Uganda. CWD-resistant clones that might not be high yielding and have low potential for cup and bean qualities and resistance to other diseases can be crossed with resistant genotypes and other lines that posses these complementary traits. The hybrid progenies can then be evaluated for all the traits starting either at the screen house through field trial testing or in field trials only. The former approach ensures that genotypes that proceed to the field are all resistant to CWD. The latter approach is less efficient for CWD resistance since it relies on field responses only, but it retains more genes especially those in moderately resistant genotypes to be tested that could have been eliminated in the more exacting screen house tests.

The current commercial clones (1s/2, 1s/3, 1s/6, 223/32, 257/53 and 258/24) used in Uganda have the complementary traits, but they are susceptible to CWD. The commercial clones yield 2.5–3 t of green coffee beans per hectare per year, have good bean qualities (18–22 g hundred beans weight, over 90% retained by screen 18/64) and have good cup qualities. Hybrids have been generated through artificial pollinations to combine the CWD resistance and the complementary traits found in these commercial clones. The progenies generated have been planted in field trials at Kituza for evaluation as individual trees. The individual genotypes that will be within acceptable limits of the traits will be selected, multiplied as clones and planted in multilocation trials in different agroecological localities for re-assessment. Clones that shall perform well in these trials will be selected for release to farmers as clonal varieties either at agroecological zone level or for the entire country depending on the performance of the variety.

Hybridization can also be between the CWD-resistant clones possessing complementary traits. The hybrid progenies generated in such crosses are also evaluated as individual trees for resistance against CWD and for field performance in the other traits. Good-performing individuals can be selected, cloned and planted in multilocation trials for adaptation and adoption tests. Good-performing clones will then be selected and released to farmers.

If there is an entire progeny of a cross between resistant and susceptible parents that perform well (resistant to CWD, CLR and red blister disease; have good qualities and are high yielding), then parents of such progenies can be planted in polycross seed gardens for production of seeds to be given out to farmers. It is anticipated that progenies involving parents from different populations, particularly from distant geographical locations, shall benefit from hybrid vigour derived from double heterozygosity of the parents. As already indicated above, given the specificity of the pathogen populations affecting the different commercial coffee species, resistant varieties can be derived through interspecific hybridizations, bearing in mind the complications associated with such hybridizations and the difficulty to derive a variety of desired quality.

As known from many breeding programmes, incorporation of CWD resistance genes into commercial clones that should also have good quality traits is likely to take a very long time, and at times, it is a gamble. Therefore, molecular techniques can be adopted to facilitate the breeding and selection process. Studies had been initiated at CORI to characterize CWD resistance using molecular markers, and it is our hope that a follow-up of these studies shall continue, which should lead to mapping of the resistance genes. Mapping studies could be initiated using the double haploid CWD-susceptible/resistant parents and their progenies to identify molecular markers and or quantitative trait loci (QTL) associated with CWD resistance to assist in breeding resistant varieties and isolation of resistance genes for creating bacterial artificial chromosome libraries. Use of double haploids shall minimize the effect of heterozygosity.

10.5.6. Grafting

Because currently the mechanism of resistance against the CWD pathogen is not known, we are not certain whether a resistant rootstock can prevent the pathogen from reaching the scion. However, if it is established that this phenomenon is possible, then grafting scions of varieties with good agronomic characteristics but which are susceptible to CWD on to a CWD-resistant rootstock should be a good and probably quicker means of deriving appropriate planting materials for farmers of both *C. canephora* and *C. arabica*. In Uganda, grafting of current commercial clones on to the CWD-resistant clones is being evaluated for (i) compatibility between the scion and rootstock and (ii) the ability of the rootstock to prevent the CWD from getting through the vascular system of the rootstock to the scion. If these results are satisfactory, then grafting should be adopted for continuing with large-scale multiplication of the current commercial varieties.

Rootstock of other *Coffea* species can also be explored for this purpose. Successful interspecific grafting involving *C. canephora* and *C. liberica* spp. has been reported (Couturon, 1993). However, Bertrand *et al.* (2001) reported depressing effects of *C. liberica* ssp. rootstock on yield and quality of *C. arabica* scion varieties. Therefore, other agronomic properties of the grafted varieties such as yield and quality should be studied and well understood before recommending grafting for producing planting materials for farmers on a large scale.

10.6. Conservation of CWD-Threatened Coffee Genetic Resources

Currently, the *C. canephora* and *C. arabica* gene pools in some Eastern and Central African countries are being depleted at varying rates by CWD. It is therefore necessary to protect the genetic resources of these species, possibly in their current diversity state. Because genetic differentiation among *C. canephora* populations is high, samples of a few genotypes with diverse genetic variability can be obtained from many populations so as to capture as much of the natural genetic variability as possible for conservation and utilization in future breeding programs. In Uganda, the *C. canephora* samples can be collected from different sites in their natural habitats such as Itwara, Kibale forests and other relic forests where wild *Coffea* exists. Samples should also be collected from isolated coffee localities such as Kalangala Islands in Lake Victoria. Because of the threat from CWD and other unforeseen natural disasters on coffee genetic resources, new locations should be identified for local *ex situ* conservation. In addition, representative samples of the genetic diversity of these species should be conserved in multiple international germplasm collections located in different countries through international collaboration.

Coffee genetic resources are threatened further by the likely adoption, on a large scale, of the incoming CWD-resistant varieties, which are likely to be derived from a few genotypes that represent a very narrow genetic base. This implies that even the few coffee plants that have withstood CWD in farmers' fields shall be abandoned and most likely destroyed as the fields are planted with the new CWD-resistant varieties. Initiation and implementation of *in situ* conservation and local germplasm utilization programmes, at regional levels within affected countries, for the remnant diversity found in devastated gardens would minimize depletion of gene pools. Such programmes are expected to sensitize the local populations and ensure that farmers multiply, conserve and beneficially utilize the plants surviving in their fields. Through such programmes, diversity at regional levels within countries can be conserved.

10.7. Conclusions

CWD has been and remains an important constraint of coffee production in Africa, and if not controlled, it could spread to affect coffee growing in other continents. Through concerted efforts, the previous epidemics in some Central and West African countries were controlled by developing and planting resistant varieties. Variety resistance is so far considered the most appropriate and effective method for controlling CWD. There has been enormous effort to develop more resistant varieties to control CWD in Uganda and Tanzania. Through these works, possible sources of intra- and interspecific resistance have been identified. In both Uganda and Tanzania, a number of resistant *C. canephora* genotypes have been identified, and Uganda in particular has already CWD clones fronted for release to farmers. Availability of these varieties will certainly help to revitalize the coffee production, not only in Uganda but also in neighbouring countries through regional cooperation, especially DRC and Tanzania. The authors hope that the highlights of challenges encountered by the breeding programmes since the re-emergence of CWD and the potential and actual procedures taken while trying to develop resistant varieties, given here, shall stand to be quick reference for future work when the same or similar outbreaks occur again, whether in Africa or elsewhere. Some challenges remain: super CWD-resistant varieties have to be checked for complementary traits and adaptation for cultivation in a wide range of geographical locations, and another challenge is the continuing depletion of the gene pools caused by CWD, whether directly or indirectly. Uganda is at the centre of diversity for *C. canephora*, so any threat to its diversity is a threat not only to Ugandan coffee production but also to the coffee sector as a whole.

Acknowledgements

We are grateful to the governments of Uganda, DRC, Tanzania and Ethiopia and our national institutes, where the studies that provided information given in this chapter were conducted, for all the support provided while these studies were being carried out. We also thank the donor communities, particularly the European Union, Common Fund for Commodities and United States Agency for International Development, for the financial support given for the studies that led to the generation of the information given in here. Last but not least, we acknowledge CAB International Africa for facilitating the writing up of this work.

References

Baradat, P. and Labbé, T. (1995) OPEP: un logiciel intégré pour l'amélioration des plantes pérennes. In: *Traitements statistiques des essais de sélection*. CIRAD, Montpellier, France, pp. 303–330.

Bertrand, B., Etienne, H. and Eskes A. (2001) Growth, production and bean quality of *Coffea arabica* as affected by interspecific grafting: consequences for rootstock breeding. In: *Proceedings of American Society for Horticulture Annual Conference No. 96*. Minneapolis, MN, pp. 206–238.

Bieysse, D. (2005) Development of long term strategy based on genetic resistance and agro ecological approaches against coffee wilt disease in Africa. EU-INCO project. Third annual report.

Couturon, E. (1993) Mise en évidence de différents niveaux d'affinité de greffes interspécifiques chez les caféiers. In: *15ème colloque de l 'Association Scientifique Internationale du Café*. Montpellier, France, pp. 209–217.

Davis, A.P., Govaerts, R., Bridson, D.M. and Stoffelen P. (2006) An annotated taxonomic conspectus of the genus *Coffea* (*Rubiaceae*). *Botanical Journal of the Linnaean Society* 152, 465–512.

Delassus, E. (1954) La trachéomycose du caféier. *Bulletin Scientifique du Ministère des Colonies, Section Agronomie Tropicale* 5, 345–348.

Figueres, R. (1940) Sur une maladie très grave du caféier en Oubangui. Rapport. Ministère des colonies, Paris, France.

Flood, J. and Brayford, D. (1997) The re-emergence of *Fusarium* wilt of coffee in Africa. In: *ASIC 1997 Proceedings of the International 17th Scientific Colloquium on Coffee Conference*. Nairobi, Kenya, pp. 621–627.

Fraselle, J. (1950) Observations préliminaire sur une trachéomycose de *Coffea robusta*. *Bulletin Agricicole, Congo Belge* XLI, 361–372.

Guillemat, J. (1946) Quelques observations sur la trachéomycose du *Coffea excelsa*. *Revue de Botanique Appliquée and d'Agriculture Tropicale* 26, 542–550.

Hakiza, G. et al. (2002) *CORI Annual Report 2001/2*.

Keuls, M. and Garretsen, F. (1977) A general method for the analysis of genetic variation in complete and incomplete diallels and North Carolina II designs. Part I: procedures and formulas from the random model. *Euphytica* 26, 537–551.

Kranz, J. and Mogk, M. (1973) *Gibberella xylarioides* Heim et Saccas on arabica coffee in Ethiopia. *Phytopathology* 78, 365–366.

Leakey, C.L.A. (1970) The improvement of robusta coffee in East Africa. In: Leakey, C.L. (ed.) *Crop Improvement in East Africa*. Government Press, Kampala, Uganda, pp. 250–277.

Lejeune, J.B.H. (1958) Rapport au Gouvernement Imperial d'Ethiopie sur la production caféière. Rapport du la FAO, Rome FAO158/3/1881.

MAAIF (Ministry of Agriculture, Animal Industry and Fisheries) (1995) Clonal Robusta Coffee Handbook. Part 1: Nursery Management Practices. Communication Centre, Entebbe, Uganda.

Meiffren, M. (1961) Contribution aux recherches sur la trachéomycose du Caféier en Cote d'Ivoire. *Café, Cacao, Thé* 5, 28–37.

Muller, R.A. (1997) Some aspects of past studies conducted in Western and Central Francophone Africa on tracheomycosis (Cote d'Ivoire, Cameroon and Central African Republic). In: Hakiza, J.G., Birikunzira, B. and Musoli, P. (eds.) *Proceedings of the first regional workshop on coffee wilt disease (Tracheomycosis)*. International Conference Centre, Kampala, Uganda, pp. 15–26.

Musoli, P.C. (2005) Breeding for resistance against coffee wilt disease. In: Bieysse, D. (ed.) *Development of long-term strategy based on genetic resistance and agro ecological approaches against coffee wilt disease in Africa*. Fourth Annual report, pp. 48–72.

Musoli, P.C. (2007) Recherche de sources de résistance à la trachéomycose du caféier *Coffea canephora* Pierre, due à *Fusarium xylarioides* Steyaert en Ouganda. PhD thesis. Montpellier II University, Montpellier, France.

Musoli, P., Olal, S., Nabaggala, A. and Kabole, C. (2001) Screening robusta coffee germplasm for resistance against coffee wilt disease. CORI progress report on coffee wilt disease research and development 1997–2000.

Pieters, R. and Van der Graaff, N.A. (1980) Resistance to *Gibberella xylarioides* in *Coffea arabica:* evaluation of screening methods and evidence for the horizontal nature of the resistance. *Netherlands Journal of Plant Pathology* 86, 37–43.

Saccas, A.M. (1951) La trachéomycose (Carbuncularise) des *Coffea excelsa,* neo-arnoldiana et robust en Oubangui-Chari. *Agronomie Tropicale* 6, 453–506.

Saccas, A.M. (1956) Recherches expérimentales sur la trachéomycose des caféières en Oubangui-Chari. *Agronomie Tropicale* 11, 7–58.

Van der Graaff, N.A. and Pieters, R. (1978) Resistance levels in *Coffea arabica* to *Gibberella xylarioides* and distribution patters of the disease. *Netherlands Journal of Plant Pathology* 84, 117–120.

11 Extension Approaches and Information Dissemination for Coffee Wilt Disease Management in Africa: Experiences From Ethiopia

E. Negussie,[1] M. Kimani,[1] A. Girma, [2]
N. Phiri[1] and D. Teshome[2]

[1]CABI Africa, UN Avenue, ICRAF Complex, PO Box 633-00621, Nairobi, Kenya
[2]Jimma Agricultural Research Center, Ethiopian Institute of Agricultural Research, PO Box 192, Jimma, Ethiopia

11.1. Introduction

Coffee plays a vital role in the economies of many African countries. However, the performance of the sector has been depressed by a number of constraining factors. In particular, coffee wilt disease (CWD) has become an issue of pressing concern in Central and Eastern African countries, including Ethiopia. Although the disease has expanded rapidly in recent years, the farming communities hold different perception and distorted views of the disease. The socio-economic baseline survey results revealed that the majority of the farmers in Central and Eastern African countries lack awareness of CWD (Chapter 6). It was reported that in Ethiopia, only 17% of the coffee farmers knew CWD as a disease but they did not know its specific name. Farmers were also not sure of the causes and ways of transmission of CWD. It was therefore realized by the Regional Coffee Wilt Programme (RCWP) that both awareness and technical knowledge on how to manage CWD were needed for farmers and other actors in the coffee sector. Consequently, awareness raising, training and dissemination of information were accorded top priority in RCWP.

There are vast numbers of Ethiopian smallholder coffee producers that operate under diverse socio-cultural, economic and natural conditions. At the same time, there is a huge disparity in terms of literacy and access to information and communication infrastructure. In addition to producers, there are several actors that are directly or indirectly engaged in and influence the coffee sector. Thus, it was well recognized that relying on a few communication channels and use of the traditional top-down and one-way approach would not effectively reach all the target groups and bring positive changes. Therefore, emphasis was placed on combining a range of complementary communication approaches and methods. These were believed to facilitate wider information dissemination and impact because most channels reinforce each other and have synergetic influences. In addition, various channels have differential impacts on different stakeholders within the coffee sector. Particularly, in view of the absence of a comprehensive package of solutions for the CWD problem, the project emphasized joint learning processes and sharing of information and experiences among relevant stakeholders. To attain this, active participation of all relevant parties was imperative. Therefore, participatory and interactive approaches were employed with farmers, extensionists and researchers who became partners and actors in the communication and learning processes.

This chapter describes the information communication strategies and approaches used by RCWP for coffee wilt management in Ethiopia.

11.2. Communication Strategies and Channels Employed

Extension education involves the conscious use of communication of information to help people learn, form sound opinions, make appropriate decisions and finally take necessary actions. Agricultural extension tries to change farmers' behaviour and practice through education and communication. 'More than natural resources, more than cheap labour, more than financial capital, knowledge is becoming the key factor of production' (World Bank, 1992). Therefore, choosing appropriate communication strategies and methods is crucial to facilitate information and knowledge sharing for change. The effectiveness of different channels varies according to the kind of information to be transferred, the target user to be addressed, the level of understanding of the receivers and other socio-economic factors. Studies show that widespread responses are seen when people are exposed to information by several methods and approaches. The communication strategy of RCWP employed both interpersonal and mass communication approaches.

11.2.1. Interpersonal communication approaches

Interpersonal communication channels involve a face-to-face exchange of information and interaction between two or more individuals. These channels promote a two-way exchange of information, facilitate immediate feedback, help to secure clarification and persuade individuals to adopt a new idea or technology. In particular, the formation and change of strongly held attitudes are best accomplished by interpersonal channels (Rogers, 1983). These channels have greater effectiveness in dealing with resistance or apathy on the part of the target groups. The principal interpersonal communication methods used by the project are described below.

Participatory hands-on training

Training sessions were the principal instruments used to create awareness and build knowledge and skills with regard to CWD management. Several training programmes were conducted for extension workers, farmers and other stakeholders. In view of the drawbacks of the traditional training approaches and the lack of complete solutions to manage CWD at present, the project employed a participatory training approach, which facilitated two-way communication and knowledge and experience sharing (Plate 33a). In contrast to the conventional training approach, which considers trainees as an empty bottle to be filled up by the trainer, the participatory approach used by the project encouraged participants to see themselves as a source of valuable information and knowledge and gave them a sense of ownership of the acquired knowledge. It allowed trainees to build up on what they know and played important role in empowering them. The training techniques and methods used are discussed under this section.

Training of trainers for extension workers

Efforts were made to train a large number of extension workers drawn from the agricultural development offices of various coffee-growing areas of Ethiopia, who in turn, train farmers and their colleagues. The trainings covered technical aspects of CWD including identification of CWD, causes, mechanisms of spread and control methods. The extensionists were also equipped with proper methods for training farmers such as adult learning principles, communication, facilitation skills and participatory training methodologies. The training sessions were characterized by active participation by the trainees and hands-on exercises. The training methodologies included interactive and brief lectures, brainstorming, buzz and large group discussions and presentations, videos of practical evidence, question-and-answer sessions, field practical, case studies and other exercises. Participants were given an opportunity to identify diseased trees, demonstrate control measures and how to practically train farmers on the field (Plate 33b). In addition, during later programmes, some of the previously trained extension workers were invited to the training sessions to share their experiences, efforts made and challenges encountered in the course of information transfer and in implementing the acquired knowledge. In total, more than 6000 extension staff received the training conducted by senior staff of the project-implementing agencies (Jimma Agricultural Research Centre and CAB International [CABI] Africa) in collaboration with the agricultural and rural development offices. The training covered five regional states, 30 zones and 93 coffee-producing districts. Extension workers trained by the project staff also played an active role in training their colleagues in the offices and those based in the field. During each training session, comprehensive handouts, posters, booklet and leaflets were prepared and given to participants. These were intended to help them remember the training messages and to effectively pass on the information to their fellow workers and farmers.

Training of farmers

At the end of each training session, the trained extension workers were encouraged to develop action plans to carry out training programmes for farmers in their respective operational areas. Through successive training programmes and various awareness creation fora, the trained extension workers made efforts to reach all farmers under their supervision. Farmers' training focused on creating awareness about the disease: its damage, identification, ways of spread, measures to control it and on the importance of mobilizing neighbours for collective action and integration of local by-laws to combat CWD. Some of the farmers' training sessions were provided by the Jimma Agricultural Research Center and CABI Africa staff directly, although most were covered by the local extension staff. In general, the trained extension workers showed remarkable commitment to reach as large number of farmers as possible through all available means. Reports communicated to Jimma Agricultural Research Centre indicated that close to half a million farmers received direct training in CWD management. Although it was not possible to obtain the exact figures, informal assessment and communication show that a considerable number of farmers and extension workers were made aware about the disease through various fora. It was also expected that there would be a trickle-down and multiplier effect of information from the trained farmers.

Training of master trainers (resource persons)

Towards the end of the project, the critical need to strongly build the capacity of selected government staff to serve as resource persons was realized. To this effect, a comprehensive training of trainers for selected coffee and crop protection experts drawn from regional, zonal and district agricultural and rural development offices was conducted. This training was different from the other 'training of trainers' courses conducted for extension workers in several aspects, including educational level of the trainees, content, intensity and length of the training course. The master trainees are expected to serve as resource persons for their respective regions and play crucial roles in the continuation of the training and information dissemination activities. They also offer advice and support on issues related to CWD. Therefore, a 10-day-long intensive and comprehensive training programme was organized for 20 participants drawn from the major coffee-growing regions. The trainees were carefully selected by considering their performance, experience and educational background. During the training workshop, a number of areas related to coffee production and protection in general and on CWD in particular were presented and discussed in detail. Moreover, relevant areas such as extension, communication, adult education, facilitation and participatory training methodologies were addressed. In addition to classroom presentations, extensive field and laboratory demonstration activities and practical exercises were carried out.

Finally, participants were encouraged to make policy analysis and recommendations in relation to CWD management. Then, they shared responsibilities to communicate these policy recommendations to their respective regions and to influence higher officials in favor of CWD management.

Participatory technology development to identify and validate CWD management options

Participatory on-farm trials were implemented for 3 years in different districts of Jimma (south-west Ethiopia), Gedeo and Sidama (Southern Ethiopia). The field trials enabled farmers to test and validate different CWD management options jointly with extension workers and researchers, as well as enhanced farmers' experimentation capacity. The steps followed in the Participatory Technology Development Process can be summarized as follows:

- Holding comprehensive stakeholders' (farmers, extension workers and researchers) training on causes, symptoms, transmission mechanisms and control measures of CWD, as well as on concepts of on-farm experiment, field trial design and implementation;
- Identification of local and scientific knowledge in relation to CWD management;
- Determining trial design and implementation (treatment application);
- Regular monitoring and data recording;
- Setting criteria and joint evaluation by farmers, extension workers and researchers;
- Sharing results through workshops, field days and exchange visits;
- Scaling out the promising ones through farmer field schools (FFS) and other mechanisms.

Based on the knowledge and experiences of local farmers, extension workers and researchers, as well as by considering experiences from other countries, seven management options (treatments) were identified during the workshop. Each treatment was applied on a plot with 15 coffee trees, and the treatments were replicated across farmers. Below are the treatments and their application.

- Use of ash: applying 2 l per tree once per annum.
- Mulch: applying once per annum at the end of the rainy reason preferably in October.
- Fungicide (copper) spray: applying once per month during the rainy season and just once every 3 months during the dry season. It is mixed at the rate of 40 g/l of water.
- Fungicide (copper) stem paint: painting the stem of a coffee tree up to 50 cm above the ground level every 4 months. 300 g of copper is added to 1 l of water.
- Herbicide (Roundup): adding 150 ml of Roundup in 15 l of water and spraying as needed based on weed condition.

- Slashing plus hand weeding: weeding by hand around the coffee tree and slashing the other parts as needed based on weed condition.
- Slashing (control): slashing the whole plot as needed.

After 3 years of running the trials, workshops were organized with the objective of reviewing the results of the field trials and to select those that can be taken to the FFS study fields. In attendance were FFS and field trial hosting farmers and facilitators, other extension staff at different levels and researchers. Researchers presented the results of the field trials during the workshop. On-farm trial hosting farmers and facilitating extension workers were also encouraged to share their observations and experiences. The participants were given an opportunity to visit the field trials and evaluate the performance and effectiveness of the different treatments. Then they were guided to set criteria against which to evaluate and choose the best CWD management options (Table 11.1).

As indicated in Table 11.1, the participants tended to choose mulch, slashing + hand weeding and ash, in that order, in southern Ethiopia. Similarly, mulch was ranked first, followed by ash and slashing + hand weeding in the Jimma area (Table 11.2). In view of the promotion of the organic coffee concept and the costs of chemical inputs, participants tended to disfavour fungicides and herbicides. In particular, mulch was ranked most highly with highest scores for all criteria except affordability, with participants talking of its additional advantages in terms of weed control, moisture and soil conservation, soil fertility improvement and avoiding slashing and other operations that would wound the coffee trees. However, in terms of affordability, use of ash was more favoured than mulch at both areas. The participants

Table 11.1. Mean scores and ranks given by 12 farmers and 14 extension staff in choosing different CWD management options in southern Ethiopia (Gedeo).

	Treatments						
Criteria	Ash	Fungicide paint	Fungicide spray	Slashing + hand weeding	Slashing (control)	Mulching	Herbicide use
Availability	4.17	1.57	1.91	4.39	4.17	4.60	2.23
Effectiveness/ efficacy	3.74	3.04	3.52	4.22	2.39	4.60	3.86
Applicability/ Simplicity	4.39	2.35	2.35	4.43	3.56	4.74	3.05
Acceptability (locally)	4.18	2.35	2.26	4.35	2.95	4.60	2.36
Affordability	4.25	2.60	2.26	4.22	3.65	3.69	2.55
Overall (average score)	4.15	2.38	2.46	4.32	3.34	4.45	2.81
Treatment rank	3	7	6	2	4	1	5

Scores: 1 = low (least preferred practice); 5 = very high (most preferred practice).

Table 11.2. Mean scores and ranks given by 17 farmers and 20 extension staff in choosing different CWD management options in south-west Ethiopia (Jimma).

	Treatments						
Criteria	Ash	Fungicide paint	Fungicide spray	Slashing + hand weeding	Slashing (control)	Mulching	Herbicide use
Availability	4.60	1.60	1.60	4.10	3.60	3.90	1.80
Effectiveness/ efficacy	3.70	3.30	3.10	3.70	1.70	4.90	3.90
Applicability/ simplicity	4.20	2.70	2.40	4.00	3.50	4.20	2.90
Acceptability (locally)	4.30	2.90	2.70	3.90	2.70	4.20	3.20
Affordability	4.40	1.90	1.70	3.50	3.10	4.10	2.10
Overall (Mean score)	4.24	2.48	2.30	3.84	2.92	4.26	2.78
Treatment rank	2	6	7	3	4	1	5

Score: 1 represents low (least preferred practice), and 5 represents very high (most preferred practice).

decided to take the selected treatments (the first three) to FFS study sites and to validate them.

After running for 3 years, some of the on-farm trials, especially those which were far away from existing FFS, were converted into FFS study sites. These helped to sustain the field trial efforts and to disseminate the preliminary findings from the field trials. It also enabled the utilization and/or sharing of the knowledge and experiences of field trial hosting farmers.

Farmer field schools

FFS have been used as an important participatory training and information dissemination tool for CWD management. FFS is a participatory training approach that can be considered both as an extension tool and a form of adult education (David *et al.*, 2006). The approach focuses on building farmers' capacity to make well-informed crop management decisions through increased knowledge and understanding of the agroecosystem. On the basis of this knowledge, farmers become independent, confident decision makers and experts in their own fields (Fliert and Van de-Fliert, 1993). The training is 'hands-on' and is carried out almost entirely in the field. The approach provides opportunity for learning by doing, through observation, discussion and interaction among participating farmers. Extension workers or trained farmers facilitate the learning process, encouraging farmers to discover key agroecological concepts and develop integrated pest management (IPM) skills through self-discovery activities practiced in the field (Ooi, 1996). More importantly, it gives an opportunity for farmers, extension workers and researchers to interact as partners in the development of IPM options. The four major principles of IPM emphasized by FFS are the following: (i) grow a healthy crop, (ii) conduct regular field observations, (iii) conserve natural enemies of crop pests and (iv) farmers become IPM experts.

Implementation of FFS under RCWP in Ethiopia

The FFS implementation concentrated on areas with high incidence of CWD such as the Gedeo, Jimma and Sidama zones. Apart from serving as a participatory learning platform, the FFS was also used as dissemination pathways for the results of the field trials. Initially, three pilot FFS were established in 2004, with a further 21 FFS groups being formed in 2005 and 2006 in southern and south-western Ethiopia. Some of the lately formed groups emerged from the on-farm field trials.

During the application of FFS approach to coffee, which is a perennial tree crop, some of the processes required changes and adaptations to the original FFS concepts to fit the local conditions and the crop under consideration. The ground work, formation and implementation of the FFS groups went through a number of processes and modifications.

Training of facilitators and curriculum development. One of the areas where modification was made is training of FFS facilitators. Initially, a 4-day intensive training workshop was held for FFS facilitators (mainly extension workers) to introduce concepts and practices of the FFS approach, adult education, group processes and management, communication and facilitation techniques. The technical training on CWD management was addressed in a separate session prior to the training of the facilitators. Moreover, technical skills were further developed in the course of actual implementation of the FFS activities. Effort was also made to enhance facilitators' knowledge and skills through provision of regular back-stopping, refresher courses and experience-sharing workshops. At the end of facilitators' training workshop, participants moved out to the field and developed a tentative curriculum for the FFS activities together with farmers, which basically follows the crop cycle or calendar. Although it primarily focused on IPM in relation to CWD, the curriculum tried to address a broad range of coffee management practices. The curriculum was flexible and regularly updated by FFS members to fit to local situations.

Community mobilization, FFS group formation and selection of study field. In general, FFS consists of groups of people with common interest who get together on a regular basis to study the 'how and why' of a particular topic (Gallagher, 2003). To establish FFS groups, the trained facilitators went back to their operational areas and held a village assembly and briefed the community about the project and aspects of CWD and explained the objectives and concepts of FFS. Then volunteers were asked to be members of the FFS group, and about 25–30 coffee farmers were registered. The selection of farmers considered interest, acceptance among the community and proximity to each other and to the study plot. Local officials, extension workers and farmers' representatives also played a role in the selection process.

One important component of FFS is the study field, which serves as a training site and an experimental laboratory where participating farmers make observations, pursue discovery-based learning and experiment with various farming techniques. One of the modifications made is in terms of plot size and crop stage. The study field, which in this case was about 0.50 ha of already established coffee farm, was provided by a group member. The criteria used in selecting the study field were accessibility to most members, proximity to the field trials (as they are expected to be a dissemination pathway for the results), trees that were relatively young and poorly managed to allow quick and clear responses to the different management practices, uniformity among trees and occurrence of CWD in the area. The selected study field was divided into two parts: coffee in one half received the farmers' conventional practices, whereas the other plot received improved management practices. This allowed farmers to compare the performance of the different management practices. The improved crop and pest management practices were determined jointly by researchers and extensionists and tested by farmers. Because of clear differences in the effects of the improved practices on the incidence of CWD, as well as on the performance of the coffee, farmers were convinced to try the improved practices/technologies on their own farms.

Regular meeting, facilitation and length of the group activity. Other areas that required adaptation was frequency of meeting and length of the group activity. In its original form developed for rice farming in Asia, FFS is a season-long activity with a weekly regular meeting. However, in view of the perennial nature of the coffee crop and its slow response to treatments, the CWD FFS were designed to operate for 2 to 3 years. Most of the groups preferred to hold their regular meeting monthly, whereas some, especially those established towards the end of the project, decided to meet fortnightly. The meetings took place in the morning between 8:30 a.m. and noon.

All RCWP FFS groups in Ethiopia have been facilitated by trained extension workers, with support from experienced farmers and researchers. It was realized that the facilitator's role and attitude are key factors in determining the success of an FFS. His or her duties include serving as catalyst, encouraging analysis, setting standards, posing questions and concerns, paying attention to group dynamics, serving as mediator and encouraging participants to ask questions and come to their own conclusion (Braun *et al.*, 2000). In short, the role of the extension workers is to facilitate the learning, experimentation and reflection processes. However, it was noticed that facilitators often tend to teach the group members in a traditional top-down way rather than encouraging them to interact, explore, discover and learn by their own. Efforts to address this problem were made through close follow-up and support by experienced project staff. Moreover, as they gained experience in running the groups and through refresher courses and interaction with other facilitators, extension workers started to improve their facilitation skills.

Why and how to conduct agroecosystem analysis?

The cornerstone of the FFS methodology is agroecosystem analysis (AESA), which involves regular and systematic observation of the crop and fields (Plate 34). FFS knowledge generation and dissemination are basically through agroecological system analysis, which is a discovery learning process. The AESA process sharpens farmers' skills in the areas of observation and decision making and helps develop their powers of critical thinking (Gallagher, 2003). During the meeting, the group was divided into small groups of five and undertook AESA. This involves field observations, analysis, discussion and presentations. Aspects of agroecosystem that were observed and analysed include growth stage of the coffee tree, occurrence and incidence of CWD and other diseases, insect pests, natural enemies, weeds, weather, soil, moisture and shade conditions. Modifications made to AESA in the CWD FFS were both in terms of frequency of observation (which was in this case fortnightly and monthly) and observation of the plant itself. Observation of some of the parts, such as the root system, was not possible so effort was made to make observation of the complete canopy. After conducting observation of the entire parts of the two plots, the groups had to randomly choose three coffee trees from each plot and make close observation, record, draw and present. The members discuss the recommendations of the small groups and take appropriate management decisions.

The importance of special topics

The special topic is the topic that participants want to learn more about. The facilitators encouraged members of FFS groups to continuously identify special topics of interest to them. The topics were then addressed either by facilitators providing information on these topics or engaging farmers in appropriate discovery learning exercises or, in most cases, inviting researchers to attend meetings to discuss the topics. Researchers provided detailed information on the identified areas and sometimes introduced various improved coffee production technologies. This particularly helped to enhance farmers' knowledge and boost adoption of improved coffee technologies. Improved coffee management practices and associated technologies applied on the improved study plot and as special topics included

- Pruning, stumping and sucker control (Plate 29);
- Shade management;
- Proper intercropping practices – coffee with haricot bean or enset;
- Proper weeding and hoeing;
- Use of mulch (Plate 30);
- Planting leguminous crops, such as *Desmodium* sp. (Plate 31);
- Compost preparation and application;
- Proper harvesting (selective picking of fully matured beans);
- Soil and moisture conservation techniques;
- Proper use of chemicals such as fungicides and herbicides.

What were the achievements and successes of CWD FFS?

The participatory, practical and flexible nature of the FFS approach was appreciated by participating farmers and created motivation and enthusiasm to seek further information and knowledge. In general, the group learning exercises enhanced farmers' awareness and knowledge about CWD. They have become experts in CWD diagnosis and are able to easily identify CWD from other diseases, such as root-rot, and tree death due to exhaustion and overbearing dieback. The process developed their critical thinking and experimental capacity. Members of the group have fully realized how improved management practices improve tree vigour and yield, and thus, they started practicing on their own farm what they have learned during the group learning activity. The process thus enhanced adoption of various improved coffee production practices. Moreover, group work helped farmers to cooperate in uprooting and burning infected coffee trees. The experience-sharing process also created interest among other neighbouring farmers to obtain new information and technologies. Thus, it has been proven that FFS groups can be promising dissemination pathways for information related to CWD in particular and for improved coffee technologies in general. Involvement in the implementation of FFS activities provided an opportunity for extension workers to develop their technical knowledge and facilitation skills. Moreover, the process created better interactions and improved linkage among farmers, extension workers and researchers.

Another significant contribution of the approach was that one of the groups in Gera district of the Jimma zone has already converted into a more permanent group that can cater for other issues. The group submitted an application to the local cooperative development office with the assistance of the facilitator. They got registered as a marketing cooperative, secured a loan and started collecting and selling members' coffee directly at the central market in Addis Ababa. During the second year, the group started purchasing other farm products in addition to coffee. Above all, the FFS process produced motivated and committed farmers who have already started making remarkable efforts to inform, teach and change other farmers. The FFS activities extend beyond the members of the FFS groups. After some suspicion in the initial stages, the FFS approach has been well received by farmers and the extension agency, and now, there is interest and demand for creation of more FFS groups.

What were the challenges and limitations?

The application of the FFS approach to tree crops such as coffee and the FFS approach in general are relatively new in Ethiopia. This somehow initially caused suspicion about the effectiveness of the approach and lack of experience in facilitating the learning process. There was a tendency among facilitators to teach farmers in a traditional top-down way, especially in the early stages, but this was gradually improved through mentoring, continuous backstopping and experience sharing. Apart from efforts to improve facilitation and technical skills of the extension workers, there was a clear need for change of attitude and mindset. In addition, because of lack of experience among facilitators, the use of group dynamics exercises and icebreakers during FFS meeting was minimal. High turnover of the trained and experienced facilitators negatively affected the group activities at some of the locations, with processes being slowed because new facilitators had to be trained and take over. There was also a tendency to prepare and present AESA predominantly in writing than in drawing, which limits involvement of illiterate farmers.

Sometimes, the process posed challenges to facilitators and researchers because farmers started coming up with certain puzzling questions and comments for which they did not have ready answers. This in fact trigged further learning, information seeking and research on the part of facilitators and researchers. Lack of appropriate coffee technologies to satisfy demands of farmers in some agroecological areas, such as the Gedeo and Sidama zones, was a major challenge. Another major challenge was meeting demand for inputs such as herbicides (to control noxious coffee weeds), seeds of cover crops such as *Desmodium* sp. (for farmers' own coffee fields) etc. Moreover, lack of effective CWD control methods such as resistant varieties, chemicals and labouriousness and ineffectiveness of the recommended uprooting and burning practices were also among the challenges encountered in the process. The perennial nature of the crop also made change and impact slow. In other words, it takes a long time to see impacts of the improved management practices. Another limitation was the low attendance of regular meetings by FFS members during harvesting seasons in some of the districts.

Awareness creation, sensitization and experience-sharing workshops and seminars

In view of the low level of awareness about CWD among different stakeholders, mass awareness creation and sensitization activities were carried out both on formally organized events and in all available informal fora. Especially, the need to create awareness and sensitize officials at different levels was frequently raised, especially by the trained extension workers. The success of the training activities and the continuity of the efforts largely depend on the attitude held and support rendered by these officials; hence, their understanding was critical. As a response to this concern, workshops were initially organized for district and zonal administrators and heads of agricultural development offices at four locations of the major coffee-growing areas. The aim of the workshop was to create awareness regarding CWD, the magnitude of the problem and consequences and to brief on overall project activities. The sessions involved introductory presentations and discussions followed by field visits to severely affected areas. In addition, national workshops were organized for higher officials, researchers, planners, policy makers and other actors. The sensitization and awareness creation activities indeed played an important role in convincing and influencing officials and policy makers. The awareness raising helped them to understand the seriousness of the problem, to put the issue of CWD at the top of their agenda and to render necessary support and commitment. For instance, coffee stumping[1] used to be implemented in a quota form, and the major concern of the extension workers was fulfiling the assigned quota by indiscriminately stumping as many coffee trees as possible. However, it is obvious that stumping activities can play a significant role in spreading CWD unless proper care and precautions are taken. As a result of the discussions on the sensitization workshops, officials and policy makers decided to stop the quota system and to undertake stumping activities with all necessary care.

In addition to the locally and nationally held workshops, researchers and project implementers attended various workshops and conferences and shared their experiences and findings with other development actors and scientists. Moreover, farmers, extension staff, researchers and other partners drawn from different countries took part in the annual regional workshops organized by RCWP. These helped to exchange views, information and experience among participants of different countries.

Awareness-raising activities were also targeted at school children because they play an important role in the coffee sector in Ethiopia. Their contribution is especially prominent in coffee weeding, harvesting and marketing. Children play a crucial role in informing and educating their parents about new developments such as incidence of diseases, new technologies, etc. Cognescent of their pivotal role, the project produced exercise books with illustrations, descriptions of the disease and control measures and were distributed to school children. To further harness this potential, agricultural teachers of different primary and secondary schools were invited to the training sessions held for extension workers. The trained teachers went back and created awareness among their students. They were also given a large number of leaflets and posters to distribute to their students. School children were also invited to take part in field days. In addition, lectures and seminars on CWD were given at higher agricultural learning institutions by researchers and senior students.

Field/open days and exchange visits

Field days were organized at some of the FFS study sites, in research stations and on on-farm field trials, and a large number of stakeholders attended the events. In total, more than 5000 stakeholders (farmers, extension staff, researchers, local officials and school children) took part in the field days organized by the project in different parts of Ethiopia. During the events, efforts were made to create an atmosphere in which visiting farmers can inspect, inquire, question, interact and get to know what has been done and the outcome. The hosting farmers played an active part in running the field days by explaining their experiences. The events offered great opportunity to exchange ideas, views and useful experiences, as well as helped to create awareness and stimulate the interests of other farmers. They also played an important role in sensitizing officials and policy makers.

[1] Stumping is a rejuvenation technique whereby the stem of an old coffee tree is cut at 30- to 45-cm height above the ground at 45° angle.

In addition, exchange visits were organized for innovative farmers, extension workers and researchers to visit other districts and regions to share experiences. The exchange visits were made to areas where there was high incidence of CWD, and this clearly demonstrated the need to pay serious attention to CWD control before huge damage has been made. Participating farmers and extension staff were able to see the severity, incidence and damage caused by the disease; how other farmers deal with the disease and coffee production systems and practices in other agroecologies. It was also noted that based on the experiences they gained from the project, different regional states and districts started organizing field days and exchange visits for their farmers, extension staff and government officials.

11.2.2. Information dissemination using printed materials and mass media

Both an old Chinese saying and modern research agree that, 'We forget what we hear, we remember what we see, but we understand what we hear, see and do'. Materials can support the talking, the seeing and the doing involved in training (Zeitlyn, 1992). The use of different print and electronic media was given due attention by RCWP in view of the difficulties of reaching all coffee-farming community only through face-to-face communication due to vastness of the coffee-growing areas, poor infrastructure and resource

Table 11.3. Differences between mass media and interpersonal communication. (Adapted from Rogers and Shoemaker, 1971, and Van den Ban and Hawkins, 1996.)

Characteristic	Interpersonal channels	Mass media channels
Message flow	Tends to be two-way	Tends to be one-way
Communication context	Face-to-face	Interposed
Amount of feedback readily available	High	Low
Ability to overcome selective processes (selective exposure)	High	Low
Speed to cover large audiences	Slow	Rapid
Possibility to adjust message to audience	Large	Small
Cost per person reached	High	Low
Possibility for audience to ignore	Low	High
Same message to all receivers	No	Yes
Possibility to persuade an individual to adopt an innovation	High	Low
Possibility to persuade an individual to form or change a strongly held attitude	High	Low
Who gives the information?	Everybody	Experts or power holders
Possible effects	Attitude formation and change	Knowledge change

limitation. Mass media can reach a large audience rapidly, spread information and create knowledge and lead to change in weakly held attitudes (Rogers, 1983). The cost per individual reached is also very low (Table 11.3). Realizing these benefits, they were used extensively to create awareness, sensitize and stimulate interest among relevant stakeholders regarding CWD.

Information dissemination through printed materials

The importance of illustrated visual aids in training and communication activities cannot be overemphasized. It was believed that the use of printed extension materials can considerably complement the other dissemination mechanisms and reinforce the communicated information. Their great advantage is that they provide a permanent and detailed record of information that can be continually referred to. People can use them in their own homes or offices and read at their own pace and whenever they want. This makes them useful and lasting reminders of the messages communicated through them. In this regard, large quantities of different types of printed materials on CWD were developed and disseminated to farmers, extension staff officials, policy makers and other stakeholders. During their production, effort was made to present the message in a comprehensible form by combining words and pictures to convey simple and clear information.

Booklets, leaflets and training handouts

Well-designed and detailed booklets were prepared in two local languages (Amharic and Oromiffa) and distributed to frontline extension workers, experts and relevant institutions. In total, more than 10,000 copies of booklets were produced and distributed by RCWP. The booklets provide detailed information on diagnosis, incidence, transmission mechanisms and management of CWD. The booklets were also given out to the trainees during the training of trainers workshops. Moreover, two types of detailed training handouts were prepared and given out to the participants of the training sessions. One of the handouts focused on various aspects of CWD, such as extent of damage, its causes, symptoms and control measures, whereas the second one dealt with adult learning, extension/communication methods and participatory training methodology. The booklets and handouts served as reference materials for extension workers during training of farmers and other field activities related to CWD.

Similarly, in view of their importance in communicating information, different types of leaflets were developed in three languages (Amharic, Oromiffa and English) and distributed to users. Leaflets are simple, informative and handy print extension materials. Leaflets and posters can be useful reminders of the spoken words (FAO, 1985). Although the leaflets were brief and prepared with simple languages, they contained complete information regarding the disease, its causes, symptoms and control measures. More than 35,000 copies of these leaflets were distributed to users in different regions, zones and districts. In addition, to those dealing with issues directly related to CWD, leaflets were also produced on coffee stumping as related to CWD. These emphasized how stumping activities spread CWD and the cares that need to be taken prior to, during and post stumping.

Posters and calendars

Posters, when they are well designed with striking images and messages that are short, catchy and to the point, can work as powerful public awareness tools (CTA, 2002). They can be used for creating awareness and to persuade and promote action. Posters can be more memorable than other print media because they are supported by pictures and diagrams. Both illiterate and educated people can be informed through simple and well-designed posters. Another advantage of posters is that once posted, they can stay around for some time and can be seen by many people again and again. Four types of posters were developed in Amharic and English, and about 11,000 copies were distributed to farmers, extension workers and different institutions. The posters were user-friendly, self-explanatory and informative with simple words and good illustrations that catch people's attention. In addition, 10,000 copies of two types of wall calendars illustrated with pictures related to the symptoms and management of CWD were produced and distributed to different stakeholders including policy makers.

Newspapers, bulletins and newsletters

Media reporters were frequently invited, and news on different events and several articles related to CWD appeared in the two national newspapers (Addis Zemen and The Ethiopian Herald), which have wide readership. These particularly played an important role in informing and sensitizing policy makers. In addition, bulletins and newsletters were also targeted to disseminate information related to CWD. Among others, the bulletin of the Ministry of Agriculture and Rural Development, Graduation bulletin of university students and quarterly newsletter of the Ethiopian Institute of Agriculture are the major ones. These channels reached a large number of people in the agricultural or coffee sector.

Other awareness creation materials

Coffee is a key commodity, and its production and use involve diverse actor groups in different ways. These stakeholders have important roles in the spread and management of the disease. Thus, with the aim of reaching these diverse groups, other awareness-creation materials such as caps, T-shirts, exercise books, stickers, etc. were produced in large quantities and were disseminated. In total, about 17,000 were produced and distributed. These materials particularly targeted producers and those involved in processing, transport, marketing, training, research and policy-makers. In particular, exercise books with illustrations and description of the disease were produced and distributed to school children.

Table 11.4. Type of mass media used and audience coverage. (From producers of respective stations/medias.)

Medium/channel	Language	Audience coverage
Radio		
Ethiopian national radio	Amharic and Oromiffa	Over 25,000,000
Radio Fana	Amharic	Over 25,000,000
Metu FM radio	Oromiffa	Over 5,000,000
SNNP FM radio	Amharic	Over 9,500,000
Ghimbi FM radio	Oromiffa	About 1,500,000
Addis FM radio	Amharic	Information not available
Television		
Ethiopian Television	Amharic, Oromiffa, English	Over 12,000,000
Newspaper and bulletin		
Addis Zemen	Amharic	Over 1,000,000
Ethiopian Herald	English	10,000 copies disseminated
MoARD Bulletin	Amharic	3,000 copies disseminated
EIAR Newsletter	English	500 copies produced; also posted on web site

Information dissemination through electronic media: radio and television

Electronic mass media are communication channels that can expose large numbers of people to the same information very quickly. Their high speed and low cost in communicating information to audiences over a wide area makes them attractive, although they are not suitable for teaching practical skills or for obtaining feedback and answering questions immediately. Information can be communicated more quickly than through printed materials or extension agents. The issue of CWD got broad media coverage (Table 11.4). News about the various events such as training, workshops, field days, cross-visits and detailed special programmes were prepared on the disease and aired to the public.

In particular, radio programmes were extensively used to effect widespread awareness creation and information dissemination on CWD, as it was felt that radio is the most widely available, accessible and affordable mass medium for remote, isolated and less literate rural communities. Radios are now common features in rural areas, even where there is no electric power. In this regard, different national and local radio stations with different languages were used to reach farmers and other stakeholders.

On the other hand, the fact that television combines vision with sound makes it attractive and influential in communicating information. Like radio, television is an instant medium that transmits information directly to a mass audience. Although in many developing countries, such as Ethiopia, television transmission and sets used to be restricted to urban areas, in recent years, they are found among some rural communities as well. Television played an important role in informing extension workers, researchers, officials and policy makers about CWD (Plate 35). It was especially instrumental in displaying the symptoms of the disease and in showing fields devastated by CWD to sensitize and motivate them to pay adequate attention to CWD management. It also helped in sharing the experiences and efforts made by different stakeholders in controlling the disease.

11.3. Summary and Conclusion

Although coffee plays a prominent role in the economies of many African countries, its productivity has been constrained by several factors. CWD has become one of the major threats facing the coffee industry in many Central and Eastern African countries. There has been general lack of awareness and knowledge about the disease among coffee farmers. This calls for efforts to enhance awareness and knowledge about CWD among different actors in the sector. Realizing this need, massive communication and information dissemination efforts were made by RCWP to reach farmers, extension workers, commercial agents, researchers and policy makers with information related to CWD. The programme used different methods, combining both interpersonal and mass communication and dissemination approaches. The former included training of trainers for extension workers, training of farmers and other stakeholders; application of FFS approach and participatory technology development; holding of various workshops and field days and exchange visits. Apart from creating awareness, these methods helped to equip farmers and extension workers with detailed knowledge and skills regarding CWD management. As they provide opportunity for face-to-face interaction and two-way exchange of information, interpersonal channels were found to be effective in persuading farmers to take necessary measures to control the disease. In particular, the FFS approach proved to be an effective tool in facilitating farmers' learning and in enhancing their critical thinking and problem-solving skills. It played an important role in creating interest among farmers to seek further information and knowledge. However, extensive effort needs to be made to improve facilitation skills of the extension workers. Continuous training, mentoring and experience sharing are needed to help them change their mindset and skills from the conventional approach to a real FFS facilitator. In addition, to further upscale and sustain the initiatives, there is a need to develop a more viable institutional framework that will provide support and ensure continuity of the FFS activities. This could involve mainstreaming FFS activities into the regular government extension system and/or forming second-order FFS associations that provide support to the groups. It is also important to encourage development of FFS groups into more permanent groups such as cooperatives or study clubs.

It was not possible to reach all stakeholders related to the coffee sector through face-to-face contact. In this regard, the use of print and electronic media played crucial role in spreading information to a wide range of stakeholders. They particularly played a vital role in sensitizing officials and policy makers about the seriousness of the problem. Among the print materials, booklets, leaflets, posters, handouts, calendars, exercise books, newsletters and newspapers were widely used. Because these media combined words, pictures and diagrams, they helped to convey clear information about the disease. They were also used as reference materials especially among extension workers. As regards electronic media, radio and television played a key role in spreading information rapidly and in creating awareness among large and diverse stakeholders.

In general, impressive results were achieved in terms of exposure and awareness creation. The use of combination of communication channels played a remarkable role in this regard. Although there have been some commendable initiatives, it is still imperative to further strengthen and secure continued support, appreciation and commitment from higher-level officials in addressing the problems of CWD. Vigorous effort should be made to accelerate information dissemination and training activities to reach all relevant stakeholders, as well as to reinforce the use and implementation of the communicated information. Farmer-to-farmer extension and information exchange through field days, exchange visits, workshops and other fora need to continue. In particular, it is important to mobilize the community to take necessary collaborative efforts because CWD management requires joint actions especially to remove infected coffee trees from the area. Organizing campaigns on CWD control in collaboration with local administrative body and other pertinent institutions is imperative. Traditional communication systems can be important channels for facilitating learning, people's participation and wider information dissemination. Therefore, their use should be seriously considered. Lack of complete and comprehensive packages of technical solutions to control CWD is among the major challenges encountered. For instance, some of the recommended control measures were found to be demanding in terms of labour or were not effective; this limited their acceptance. Therefore, there is a dire need to strengthen research efforts to generate appropriate and viable CWD management options that can be communicated to end-users.

Acknowledgements

This chapter is an output of the Project 4 activities of RCWP, which was financed by the Common Fund for Commodities. The International Coffee Organization supervised the project, whereas CABI was the project execution agency. In Ethiopia, the project was implemented by the Jimma Agricultural Research Centre of the Ethiopian Institute of Agricultural Research. The authors sincerely thank all the institutions that participated in implementing the project activities. The authors also wish to express their gratitude to Dr Dannie Romney for her valuable comments on the draft article.

References

Braun, A.R., Thiele, G. and Fernandez, M. (2000). FFS and Local Agricultural Research Committees: Complementary Platforms for Integrated Decision-making in Sustainable Agriculture. AgREN Network paper No. 105. ODI.

CTA (2002) Public awareness: a manual for agricultural NGOs and research institutions in Africa.

David, S., Agordorku, S., Bassanaga, S., Couloud, J., Kumi, M., Okuku, I. and Wandji, D. (2006) *A Guide for Conducting FFS on Cocoa Integrated Crop and Pest Management.* International Institute of Tropical Agriculture, Accra, Ghana.

FAO (1985). Guide to Extension Training. FAO Training Series. Rome.

Fliert, E.V.D. and Van de-Fliert, E (1993) Integrated pest management: FFS generate sustainable practices. *Wageningen Agricultural University Papers* 34, 177–184.

Gallagher, K. (2003) Fundamental elements of a farmer field school. *Magazine on Low External Input and Sustainable Agriculture* 19, pp. 5–6.

Ooi, P. A. C. (1996) Experiences in educating rice farmers to understand biological control. *Entomophaga* 41, pp. 375–385.

Rogers, E. M, (1983). *Diffusion of Innovations,* 3rd edn. Free Press, New York.

Rogers, E.M. and Shoemaker, F.F. (1971) *Communication of innovations.* Free Press, New York.

Van den Ban, A.W. and Hawkins, H.S. (1996) *Agricultural Extension,* 2nd edn. Blackwell Science Ltd, Oxford, UK.

World Bank (1992) Policy Research Bulletin. Washington, DC. World Bank.

Zeitlyn, J., (1992) Appropriate Media for Training and Development. University Press Ltd, Dhaka, Bangladesh.

12 Concluding Remarks

Julie Flood
CABI, Bakeham Lane, Egham, Surrey, TW209TY

From the proceeding chapters, it is obvious that much has been achieved during the Regional Coffee Wilt Programme (RCWP). We know more about the biology of the pathogen, the extent of its spread and its impact on the livelihoods of smallholder producers across participating countries than we did before the RCWP started. Producers and other stakeholders in the coffee sector in participating countries are now better equipped to recognize the disease so that sanitation measures can be conducted quickly. We also know that breeding for resistance remains a very good option for management of this disease (an approach used successfully in earlier epidemics in the 20th century). All this information will certainly help the coffee sector in countries currently affected by the disease. It will also help to alert other countries to the threat of coffee wilt disease (CWD). We have learned much, but we also should consider what lessons can be learned for the future.

It has been estimated that around 1 billion dollars has been lost from this disease due to both direct losses from reduced coffee production and in costs incurred to try to manage the disease. Affected countries have been changed forever, with many farmers moving away from coffee production completely – either diversifying into other crops or moving out of agriculture altogether. What can we learn from the outbreak of CWD as an example of a major epidemic of such a commercially important crop as coffee and a crop so extensively grown in Africa? Essentially, the enormous impact it has had across the continent was largely due to a lack of preparedness for a major epidemic caused by a pest/disease in coffee-producing countries in Africa. This was despite the fact that just a few decades previously, the disease had reached epidemic proportions in several African coffee-producing countries and had caused major effects on production and on coffee diversity. Inadequate preparedness was compounded by failures to understand the seriousness of the threat and failures to respond quickly with counteractive measures. National governments were initially slow to recognize the threat to their economies and to producer livelihoods, and this was exacerbated by the failure of the international community to recognize the seriousness of the situation at an early stage. CWD was considered to be a minor problem and one that was already under control. Undoubtedly, if there had been fewer delays, the disease could have been managed more effectively and its spread limited. Because of the remote location of many of the initial disease foci, many years elapsed from the start of the new epidemic in the Democratic Republic of Congo (DRC; probably commencing in the 1970s) to the international community being aware of the seriousness of the situation in that country and the likely spread to other countries. Uganda and adjacent countries were not alerted to the seriousness of the situation until it was too late.

Even when the international community was alerted, delays continued with many years elapsing from the initial surveys conducted in the 1990s to project activities being implemented in the field in the early 2000s. This illustrates the point that international funding agencies can be slow to respond to issues relating to plant health, with far-reaching consequences. For diseases such as swine flu or avian flu, which are seen to directly affect human health, resources can be mobilized within days or weeks, but it can take years to mobilize funds to manage an epidemic plant disease despite the direct effect on the livelihoods of hundreds of thousands of small producers. In the intervening years, the disease had become widespread in Uganda and had emerged in Tanzania. In retrospect, although a programme such as the RCWP was clearly urgently needed for the long-term understanding of the pathogen, to raise awareness across the coffee sector and to initiate breeding programmes for the sustainable management of CWD, there should also have been a mechanism for developing countries to obtain international funds quickly in order to respond much more rapidly and restrict disease spread. This would have involved alerting coffee stakeholders in affected countries and adjacent countries and conducting emergency sanitation measures. Some national governments did not wait for intervention from the international community. Given the gravity of their situation, a strategy was developed in Uganda using internal funds. Further surveys were conducted to ascertain the full extent of the problem, and a programme of awareness raising for farmers and a research strategy, including a breeding programme, were initiated. Uganda was fortunate enough to have a good research base and agricultural infrastructure to be able to implement these activities. Similarly, Tanzania initiated surveys and alerted producers in robusta-growing areas adjacent to Uganda – rapid action that helped to reduce the impact of CWD in the country to some extent. These activities undoubtedly received a much deserved boost from international funds channelled through the RCWP, and progress has been much more rapid. Participating countries also benefited from institutional capacity building both through training programmes and improvements in infrastructure. International and regional expertise were mobilized through the RCWP. In the case of Uganda, lessons were learned from CWD as can be seen from the speed of the national response to banana bacterial wilt that emerged in Mukono in 2002. Farmers were mobilized quickly to conduct sanitation, and a research agenda was identified and funded internally by the national government, all within 1 year of the initial report of the disease.

Nevertheless, despite the successes, CWD remains endemic to many parts of central and eastern Africa and is still spreading. To try to prevent the same problems arising again in adjacent countries, regular surveys should be undertaken and the coffee sectors in these countries should be alerted of the threat to their industry. We need to improve national capacity to recognize and deal with threats like CWD and encourage adjacent countries to put a strategy in place to facilitate this. Because early recognition of CWD symptoms is a crucial part of successful management, making authorities and producers in neighbouring countries aware of the presence of the disease and its imminent threat is essential. The disease is spreading in the Equateur Province in DRC, putting robusta coffee production there under threat, as well as robusta production in adjacent countries such as Congo, Central African Republic, Gabon and Cameroun. In many remote areas, traffic of people and goods including coffee, across porous borders is likely to add to the risk of further disease spread. There is a greater threat when mass movements of people take place in areas of conflict or due to natural disasters. Spread of plant disease is often ignored in those circumstances and yet can have far-reaching consequences, as has been highlighted here. The conflict that erupted in eastern DRC during the 1990s undoubtedly contributed to the spread of CWD on robusta coffee. In addition, isolates of the pathogen from arabica coffee in Ethiopia are pathogenic to arabica coffees grown in Kenya, and the Kenyan coffee sector needs to be aware of this. Indeed, this disease is not just a threat to African coffee production. Coffee from other producing countries has now been shown to be susceptible, so the disease is a threat to coffee production globally and coffee producers worldwide should be made more aware of this disease. It is the aim of the many contributors to this book, including myself, that it should be one of the avenues of communication by which the global coffee industry is alerted to CWD and the threat it presents.

The CWD pathogen also infects wild coffees. As such, it constitutes a threat to coffee diversity, and this has significant implications for coffee improvement programmes worldwide. The disease is widely distributed in Ethiopia, an important source of diversity for arabica coffee and where wild forest arabica coffee is still found in the south-west of the country. Equally, in Uganda, where relic forests exist, wild *Coffea canephora* can be found, e.g. in Kibale forest. Wild *C. canephora* also exists in DRC. The existence of this pathogen over large areas of the continent, which correspond to the centres of diversity for both robusta and arabica coffees, could have a huge impact on future coffee breeding globally because genetic diversity is reduced. Such a reduction in diversity will not only restrict our ability to breed for pest and disease resistance but will also impact on the improvement of many agronomic traits. The destruction of coffee germplasm collections during this current epidemic has also affected national and regional strategies to manage pests and diseases, as well as improvements for other agronomic traits. In short, the disease has reduced the diversity of material available for future breeding programmes. Paradoxically, with the selection of CWD-resistant material in-country, this effect may be further compounded, as farmers may opt to use resistant material extensively, further reducing diversity of material available for breeding programmes. Conserving the remaining coffee diversity both in-country and in international collections should be an area for future work.

Yet, one of the notable successes of the RCWP has been the identification and selection of resistant material in several countries. Multiplication of this material is now needed for dissemination to growers. Further funding is needed to support this, as it will necessitate mass multiplication of genotypes currently considered as resistant to CWD and distribution of this material to farmers. Finance will also be required to support future coffee-breeding programmes and coffee breeders. The situation highlights limitations relating to the cycle of project funding and how progress made during one project can be sustained into the future when investment is removed. We do not yet know how long this germplasm will remain resistant to the disease in the field. Improvement programmes should be ongoing. The disease is likely to spread further in Africa, and resistant germplasm will be needed in more countries. Strengthening coffee institutions and the coffee sector is needed to enhance competitiveness, and consequently, long-term investment for revitalizing the African coffee sector is required. Currently, the coffee sector in Africa is characterized by low productivity and an aging workforce who need to know more about improved production methods, good agricultural practices and improving their market access. Investment is required both from national governments, so that their policies reflect support for the sector, and from other sources, including the private industry and international funding agencies. In many respects, the decline in investment to the coffee sector reflects a general lack of investment in agriculture, which has taken place over many decades. Such a decline needs to be reversed. The net result of decades of under-investment is a weakened agricultural infrastructure and weakened institutions in many developing countries. The re-emergence of CWD as a major production constraint for coffee production in several African countries is an excellent illustration of what can go wrong due to the many complex problems associated with this decline in agricultural investment that has taken place since the 1980s.

October 2009

Index